First steps *in* Modal Logic

First steps
in
Modal Logic

Sally Popkorn

CAMBRIDGE
UNIVERSITY PRESS

CAMBRIDGE UNIVERSITY PRESS
Cambridge, New York, Melbourne, Madrid, Cape Town, Singapore, São Paulo

Cambridge University Press
The Edinburgh Building, Cambridge CB2 8RU, UK

Published in the United States of America by Cambridge University Press, New York

www.cambridge.org
Information on this title: www.cambridge.org/9780521464826

First published 1994
This digitally printed version 2008

A catalogue record for this publication is available from the British Library

ISBN 978-0-521-46482-6 hardback
ISBN 978-0-521-05793-6 paperback

Contents

Introduction   xi

Acknowledgements   xiii

I   Preliminaries   1

1   Survey of propositional logic   3
   1.1   Introduction 3
   1.2   The language 4
   1.3   Two-valued semantics 6
   1.4   The proof theory 7
   1.5   Completeness 9
   1.6   Exercises 11

2   The modal language   13
   2.1   Introduction 13
   2.2   The language defined 15
   2.3   Some particular formulas 16
   2.4   Substitution 17
   2.5   Two remarks 19
   2.6   Exercises 19

II   Transition structures and semantics   23

3   Labelled transition structures   25
   3.1   Introduction 25
   3.2   Some examples 27
   3.3   Modal algebras 28
   3.4   Various correspondences 30
   3.5   The diamond operation 32
   3.6   The structure regained 34
   3.7   Exercises 35

4 Valuation and satisfaction   **39**
  4.1 Introduction 39
  4.2 The basic satisfaction relation 41
  4.3 Some examples 42
  4.4 The three satisfaction relations 46
  4.5 Semantics for modal algebras 48
  4.6 Exercises 51

5 Correspondence theory   **61**
  5.1 Introduction 61
  5.2 Some examples 61
  5.3 The confluence property 64
  5.4 Some non-confluence properties 66
  5.5 Exercises 69

6 The general confluence result   **77**
  6.1 Introduction 77
  6.2 The structural property 79
  6.3 The set of formulas 82
  6.4 The result 83
  6.5 Exercises 84

III Proof theory and completeness   **89**

7 Some consequence relations   **91**
  7.1 Introduction 91
  7.2 Semantic consequence 92
  7.3 The problem 93
  7.4 Exercises 94

8 Standard formal systems   **97**
  8.1 Introduction 97
  8.2 Formal systems defined 98
  8.3 Some monomodal systems 102
  8.4 Some polymodal systems 106
  8.5 Soundness properties 108
  8.6 Exercises 110

9 The general completeness result   **119**
  9.1 Introduction 119
  9.2 Statement of the result 120
  9.3 Maximally consistent sets 121
  9.4 The canonical structure 124
  9.5 The canonical valuation 125

9.6   Proof of the result 126
9.7   Concluding remarks 126
9.8   Exercises 127

10 Kripke-completeness   **129**
10.1 Introduction 129
10.2 Some canonical systems 130
10.3 Confluence induced completeness 133
10.4 Exercises 137

IV   Model constructions   **139**

11 Bisimulations   **141**
11.1 Introduction 141
11.2 Zigzag morphisms 142
11.3 Bisimulations 144
11.4 The largest bisimulation 147
11.5 A hierarchy of matchings 148
11.6 An example 150
11.7 Stratified semantic equivalence 151
11.8 Exercises 154

12 Filtrations   **157**
12.1 Introduction 157
12.2 The canonical carrying set 159
12.3 The left-most filtration 160
12.4 The right-most filtration 161
12.5 Filtrations sandwiched 163
12.6 Separated structures 164
12.7 Exercises 166

13 The finite model property   **169**
13.1 Introduction 169
13.2 The fmp explained 169
13.3 The classic systems have the fmp 172
13.4 The basic temporal system has the fmp 180
13.5 Exercises 182

V   More advanced material   **185**

14 SLL logic   **187**
14.1 Introduction 187
14.2 The *-closure of a relation 189

14.3 The axioms for SLL 189
14.4 SLL is not canonical 191
14.5 A filtration construction 193
14.6 The completeness result 197
14.7 Exercises 197

15 Löb logic   201
   15.1 Introduction 201
   15.2 The system defined 206
   15.3 The rule of disjunction 206
   15.4 The fmp 209
   15.5 Exercises 214

16 Canonicity without the fmp   217
   16.1 Introduction 217
   16.2 The system defined 217
   16.3 The characteristic properties 218
   16.4 Canonicity 219
   16.5 The finite models 220
   16.6 A particular model 222
   16.7 Exercises 222

17 Transition structures aren't enough   225
   17.1 Introduction 225
   17.2 The system KZ 226
   17.3 The system KY 227
   17.4 A particular structure 228
   17.5 Exercises 232

VI Two appendices   235

A The what, why, where,... of modal logic   237
   A.1 Introduction 237
   A.2 Beginning 237
   A.3 About this book 239
   A.4 What next? 241
   A.5 Some uses of modal logic 242

B Some solutions to the exercises   247
   B.1 Chapter 1 247
   B.2 Chapter 2 250
   B.3 Chapter 3 252
   B.4 Chapter 4 254
   B.5 Chapter 5 260

CONTENTS

B.6   Chapter 6 266
B.7   Chapter 7 267
B.8   Chapter 8 267
B.9   Chapter 9 278
B.10 Chapter 10 280
B.11 Chapter 11 286
B.12 Chapter 12 289
B.13 Chapter 13 293
B.14 Chapter 14 299
B.15 Chapter 15 304
B.16 Chapter 16 306
B.17 Chapter 17 308

Bibliography   311

Introduction

This book is an introduction to modal logic, more precisely, to classically based propositional modal logic. There are few books on this subject and even fewer books worth looking at. None of these give an acceptable mathematically correct account of the subject. This book is a first attempt to fill that gap.

Apart from its mathematical clarity, some other features of the book are:

- The central concept of the book is that of a labelled transition structure, and polymodal languages are used from the beginning.

- Modal languages are viewed as a tool for analysing the properties of transition structures, not the other way round.

- There is not an overemphasis on syntactic (proof theoretic) matters.

- Nevertheless, a detailed explanation is given of the differences between the weak completeness and Kripke completeness of formal systems.

- Correspondence properties (the expressibility properties of modal languages) are stressed as an important tool.

- Bisimulations are used as a method of comparing transition structures.

- Each chapter has a decent selection of exercises and over one sixth of the book consists of a comprehensive set of solutions to these exercises.

The book is aimed primarily at a computer science readership. However there is no computer science in the book and very little material which is directly attributable to a computer science motivation. Thus the reader of the book may be interested in modal logic in its own right or because of one or several of its applications in computer science. To read the book it is not necessary to understand any of these fields of application in any great depth.

The applications and uses of modal logic are many and varied. Aspects of the subject can be found in

- The analysis of tenses

- Concurrency

- Belief logic and default logic

- Program correctness

- Power domain constructions

- Situation theory

and several other areas. To emphasize some of these in prefence to others would only restrict the possible readership. The aim of the book is simply to give a correct and concise account of the core of the subject with just a hint of the more advanced topics. The aim is not to describe the possible applications of the subject.

In order to keep the book to a reasonable size there are some important topics which the book does not attempt to cover. The book covers only propositional modal logic on a classical base. It does not consider any predicate logic versions of modal logic, nor does it consider any base logic other than classical 2-valued logic. The book does not give a detailed discussion of the various proof theoretic ramifications of modal logic. Only the simplest and most routine proof system is discussed. Nor is it concerned with the execution or implementation problems of model systems. Almost nothing is said about decision procedures for modal logic.

No doubt there are also other topics which you would have liked included, however these are for a more advanced or specialized book.

For convenience the book is divided into six reasonably sized parts. Each of these is devoted to one aspect of the subject as follows.

PART I, which covers Chapters 1 and 2, gives a survey of the required material concerning propositional logic and then introduces the family of (propositional) modal languages.

PART II, which covers Chapters 3, 4, 5, and 6, is concerned with various semantic matters. Chapter 3 introduces the structures which support the Kripke style semantics for modal languages, and Chapter 4 discusses these various semantics. The two Chapters 5 and 6 give various results which illustrate how properties of the supporting structures can be captured by modal formulas.

PART III, which covers Chapters 7, 8, 9, and 10, is concerned with proof theoretic matters. Firstly, in Chapter 7 various motivating semantic consequence relations are discussed. Then in Chapter 8, the standard proof theoretic machinery (as used in this book) is introduced and developed as far as is needed. Finally, in Chapters 9 and 10, two completeness results are proved. These show how the semantics and the proof theory interact in a nice way.

These three parts, I, II, and III, form (what should be) the basis of a first course in modal logic.

PART IV, which covers Chapters 11, 12, and 13, deals with more advanced, but still fairly basic, material. In particular, the notion of a bisimulation, the construction of models using filtrations, and the finite model property of certain formal systems are discussed.

PART V, which covers Chapters 14, 15, 16, and 17, gives four examples of more advanced, but interesting, topics. The first two examples (in Chapters 14 and 15) illustrate some of the power of modal logic, and the second two examples (in Chapters 16 and 17) illustrate some of the problems which come with this power.

PART VI consists of two Appendices. The first of these, Appendix A, contains both the first and last things you should read in the book. It is a general discussion of modal logic, from the various other introductory texts, what you can expect from this text, what you should read next, and a survey of some of the uses of modal logic. As soon as you finish this Introduction, i.e. at the end of the next sentence, you should go immediately to this Appendix. Finally, Appendix B is a fairly comprehensive set of solutions to the exercises given at the end of each chapter.

Acknowledgements

Frances Nex picked up several mistakes in early drafts of the book, and made many suggestions that improved the layout.

David Murphy kept me straight on some of the LaTeX eccentricities.

Paul Taylor's customized package was used to produce the diagrams in this book.

However, anything between the covers that doesn't please you is entirely to my credit. The cover itself is the work of the C.U.P.

On a personal note I am grateful to Francis Lee, Joseph Holt, and Raquel who, at the time of writing, promise a better future, stand for traditional values, and live in a different world.

Saturday of the Lord's Test
July 1994

Part I

Preliminaries

This part, which consists of two chapters, gives the basic material. Chapter 1 is a brief discussion of the required background from (modal-free) propositional logic. There should be nothing in this chapter which you don't know already, although perhaps the style of presentation will be new to you. The variety of modal languages are described in Chapter 2. The polymodal versions are introduced right from the beginning and this brings a greater cohesion to the subject.

Chapter 1

Survey of propositional logic

1.1 Introduction

Propositional logic is an analysis of the natural language connectives

not

if ... then...

and

or

if ... and only if ...

⋮

as used in a certain restricted context. Thus the analysis is not intended to cover all possible uses of these words in natural language, but only those uses in 'logical arguments' where the meanings of the words are determined in a truth-functional way.

In order to make this context clear the analysis is undertaken via the medium of an abstract, but precisely defined, formal language, the *propositional language*.

The first part of the analysis, the semantics, shows how a truth value can be ascribed to sentences of this language, and then makes precise the notion of the 'logical consequences' of a set of such sentences. This part of the analysis makes use of a standard semantics, i.e. it makes reference to the intended meanings of the connecting symbols of the language (which, of course, are the connectives *not*, *if...then...*, *and*, ...).

The second part of the analysis, *propositional calculus*, shows how the notion of 'logical consequence' can be simulated by certain combinatorial manipulations within the language. This is done entirely abstractly without reference to any intended meaning. This simulation can be done in several different ways each making use of a different style of *formal system*. (For classical propositional logic, which is what we are concerned with here, the differences between these styles are more a matter of taste than content.)

3

The culmination of the analysis is a proof of *completeness*. The chosen formal system is first shown to be *sound* in that anything which is simulated as a logical consequence is one; and then it is shown to be *adequate* in that every logical consequence can be simulated within it.

I assume that, to some extent, you are already familiar with this material. If you are not then you shouldn't be reading this book; there is no point in trying to learn modal logic unless you have a firm grasp of the underlying propositional logic. If you do not have this background I suggest you first acquire it from one (or several) of the many available textbooks covering the subject (some of which are quite good).

In this chapter I will give a brief survey of classical, 2-valued, propositional logic in a form suitable for extension to the modal case. There are many different styles of systems of propositional calculus (Hilbert, Natural, Sequence, ...) all having their good and their bad points. We are not concerned with these pros and cons here; in particular we are not concerned with proof theoretic efficiency (even though this is an important topic which must be addressed eventually). This book is an *introduction* to modal logic, and as such it will present an overview of the basics of the subject rather than the intricacies of the more detailed analysis of certain of its aspects or fields of application.

1.2   The language

So let us begin the refresher course.

The first thing we do is define the abstract, but precisely constructed, *propositional language*. This is built up from certain *primitive symbols* comprising the *variables*, the *connectives*, and the *punctuation symbols*. These are combined in certain ways to produce the *formulas*. The connectives are intended to represent the English language connectives not, if...then..., etc. Since connectives need something to connect, the variables provide a starting point for the process. The punctuation symbols are precisely that; they are used to ensure that the formulas are uniquely readable.

The primitive symbols of the language are:

- The elements $P, Q, R, \ldots$ of a fixed countable set Var of *variables*;

- The *propositional connectives*

$$\top, \bot, \neg, \rightarrow, \wedge, \vee$$

    of 0, 0, 1, 2, 2, and 2 argument places, respectively;

- The *punctuation symbols* (and).

The formulas of the language are constructed in the usual way.

1.1 DEFINITION. The formulas of the language are obtained recursively using the following clauses.

(atomic) Each variable $P \in Var$ and each constant $\top$ and $\bot$ is a formula.

(propositional) For all formulas θ, ψ, ϕ each of

$$\neg\phi \quad , \quad (\theta \to \psi) \quad , \quad (\theta \wedge \psi) \quad , \quad (\theta \vee \psi)$$

is a formula.

Let $Form$ be the set of all formulas. ∎

The countability of Var is a restriction on the size of the set. If you know what this means then you will recognize where it is used later. If you do not know what it means then, for the purposes of this book, you may regard Var as a given by a list

$$P_0, P_1, P_2, P_3, \cdots, P_r, \cdots$$

However, sometime in the future you should find out what the word means, and how it effects some of the arguments later on.

Note that formulas are defined by a recursion procedure. This means that some facts about formulas can be proved by *structural induction*, i.e. by an induction on the structure of formulas.

For instance, suppose Φ is some set of finite strings of primitive symbols and suppose we know the following.

(0) Φ contains all variables and the two constants $\top$ and $\bot$.

($\neg$) For all formulas θ

$$\theta \in \Phi \quad \Rightarrow \quad \neg\theta \in \Phi.$$

($*$) For all formulas θ and ψ

$$\theta, \psi \in \Phi \quad \Rightarrow \quad (\theta * \psi) \in \Phi$$

(for each binary connective $*$).

We may then conclude that Φ contains all formulas. For suppose not, i.e. suppose there is at least one formula with $\phi \notin \Phi$. Consider an example of such a ϕ containing the least number of symbols. This ϕ can not be a variable or constant, by (0). It must, therefore, have the shape

$$\neg\theta \quad \text{or} \quad (\theta * \psi)$$

for some formulas θ and ψ and connective $*$. But both of these lead to contradictions, by ($\neg$) or ($*$). Thus our original assumption is wrong, hence there is no formula which is not in Φ.

When displaying particular formulas we sometimes omit various brackets and use various other devices to aid readability. However, these displayed strings are not themselves formulas (but just pictures of formulas).

·1.3 Two-valued semantics

Let $2 = \{0, 1\}$ and think of 2 as the 'truth object'. We regard 0 as FALSE and 1 as TRUE. Each connective has an associated operation on 2. The operation

$$\neg : 2 \to 2$$

associated with the connective $\neg$ is given by

$$\neg(x) = 1 - x$$

(for each $x \in 2$). Each binary connective $*$ has an associated operation

$$* : 2 \times 2 \to 2$$

given by the following truth table.

		$x * y$		
x	y	$\to$	$\wedge$	$\vee$
0	0	1	0	0
0	1	1	0	1
1	0	0	0	1
1	1	1	1	1

Notice how this defines the intended meaning of the symbols

$$\neg \qquad \to \qquad \wedge \quad \vee$$

as

$$\text{not} \quad \text{if ... then} \quad \text{and} \quad \text{or.}$$

(Note also that we are using the same symbol for the formal connective and its operational counterpart on 2. This should not lead to confusion.)

The basic semantic notion is the construction of the truth value of a formula ϕ. This can not be done in a vacuum, but only within a context where the truth values of the variables are known. The whole process is encapsulated as follows.

1.2 DEFINITION. A *valuation* is a map

$$\nu : Var \longrightarrow 2.$$

For each such valuation ν the associated map

$$[\![\cdot]\!]_\nu : Form \longrightarrow 2$$

is defined by recursion on the structure of formulas using the following clauses.

(Const) For the constants

$$[\![\top]\!] = 1 \quad , \quad [\![\bot]\!] = 0.$$

(**Var**) For each variable P

$$[P] = \nu(P).$$

($\neg$) For each formula θ

$$[\neg\theta] = 1 - [\theta].$$

($*$) For all formulas θ, ψ

$$[(\theta * \psi)] = [\theta] * [\psi]$$

(for an arbitrary binary connective $*$). ■

(As in this definition, when using $[\cdot]_\nu$, it is usual to drop the distinguishing subscript ν unless this could lead to confusion.)

We say a valuation ν *models* or is a *model* of a formula ϕ, or that ϕ *is true for ν* if

$$[\phi] = 1.$$

We can now make precise the notion of 'logical consequence'. Thus, given a set Φ of formulas and a formula ϕ

$$\Phi \models \phi$$

means that ϕ is true for every model of (all members of) Φ. When this holds we say ϕ is a *semantic consequence* of Φ. Formulas ϕ such that

$$\models \phi$$

(i.e. which are true for all valuations) are called *tautologies*.

1.4 The proof theory

The objective of propositional calculus is to give a syntactic description of the semantic consequence relation $\models$ by setting up an appropriate formal system. This can be done in many different ways; here we describe a system that is the most convenient for later generalization to the modal situation. We describe a system in the Hilbert style.

Thus we first set down the set of *logical axioms*. These will be tautologies and typically will contain all formulas of the shapes

$$(k) \quad \phi \to (\theta \to \phi)$$
$$(s) \quad \theta \to (\psi \to \phi) . \to . (\theta \to \psi) \to (\theta \to \phi)$$

together with enough axioms to control the other connectives. We also use just one *rule of inference*, modus ponens.

$$(\text{MP}) \ \frac{\theta \quad \theta \to \phi}{\phi}$$

These are used to generate the proof theoretic consequence relation $\vdash$.

(It is important to notice that we are dealing with the connectives via the use of axioms and not extra rules of inference. In the propositional case there is a relatively easy way to trade off the use of axioms against the use of rules of inference, however in the modal case this is not so easy, so we base our system on just the one rule.)

1.3 DEFINITION. Let Φ be an arbitrary set of formulas.

(a) A *witnessing deduction* from Φ is a sequence

$$\phi_0, \phi_1, \cdots, \phi_n$$

of formulas such that for each formula ϕ_i of the sequence, at least one of the following holds.

(hyp) $\phi_i \in \Phi$.

(ax) ϕ_i is a logical axiom.

(mp) There are formulas ϕ_j, ϕ_k occurring earlier in the sequence (i.e. with $j, k < i$) such that $\phi_k = (\phi_j \to \phi_i)$.

(b) For each formula ϕ, the relation

$$\Phi \vdash \phi$$

holds precisely when there is a witnessing deduction from Φ with ϕ as the last term of this deuction. ∎

This relation

$$\Phi \vdash \phi$$

is the simulation of the notion of logical consequence.

Recall that this formal system has the Deduction Property, that is for each set of formulas Φ and pair of formulas θ, ϕ the implication

$$\Phi, \theta \vdash \phi \quad \Rightarrow \quad \Phi \vdash (\theta \to \phi)$$

holds. This is an important property which fails to hold for most modal systems.

1.5 Completeness

It is straight forward to show that the formal system is sound, i.e. that

$$\Phi \vdash \phi \quad \Rightarrow \quad \Phi \models \phi.$$

This is proved by a routine induction on the length of the witnessing formal deduction.

The proof of adequacy (and hence completeness) takes a little longer and can be achieved in several different ways. Here I will sketch a proof which later will form the basis of the corresponding proof for modal systems.

We say a set of formulas Φ is *consistent* if

$$\text{not}[\Phi \vdash \perp].$$

Let **CON** be the set of all such consistent sets Φ. The formal system is designed to achieve the following properties of **CON**.

(Finite character) For each set of formulas Φ we have $\Phi \in \mathbf{CON}$ precisely when $\Psi \in \mathbf{CON}$ for each finite $\Psi \subseteq \Phi$.

(Basic consistency) For each variable P we have $\{P, \neg P\} \notin \mathbf{CON}$ and, of course, $\{\perp\} \notin \mathbf{CON}$.

(Conjunctive preservation) For all appropriate θ, ϕ and Φ with $\Phi \in \mathbf{CON}$

$$\begin{aligned}
(\theta \wedge \psi) \in \Phi &\Rightarrow \Phi \cup \{\theta, \psi\} \in \mathbf{CON} \\
\neg(\theta \vee \psi) \in \Phi &\Rightarrow \Phi \cup \{\neg\theta, \neg\psi\} \in \mathbf{CON} \\
\neg(\theta \rightarrow \psi) \in \Phi &\Rightarrow \Phi \cup \{\theta, \neg\psi\} \in \mathbf{CON}.
\end{aligned}$$

(Disjunctive preservation) For all appropriate θ, ϕ and Φ with $\Phi \in \mathbf{CON}$

$$\begin{aligned}
(\theta \vee \psi) \in \Phi &\Rightarrow \Phi \cup \{\theta\} \in \mathbf{CON} \quad \text{or} \quad \Phi \cup \{\psi\} \in \mathbf{CON} \\
\neg(\theta \wedge \psi) \in \Phi &\Rightarrow \Phi \cup \{\neg\theta\} \in \mathbf{CON} \quad \text{or} \quad \Phi \cup \{\neg\psi\} \in \mathbf{CON} \\
(\theta \rightarrow \psi) \in \Phi &\Rightarrow \Phi \cup \{\neg\theta\} \in \mathbf{CON} \quad \text{or} \quad \Phi \cup \{\psi\} \in \mathbf{CON}.
\end{aligned}$$

(Negation preserving) For all appropriate θ and Φ

$$\neg\neg\theta \in \Phi \in \mathbf{CON} \quad \Rightarrow \quad \Phi \cup \{\theta\} \in \mathbf{CON}.$$

Now let $\boldsymbol{S}$ be the set of all the maximally consistent sets of formulas, i.e. those $\Phi \in \mathbf{CON}$ such that for all sets Ψ

$$\Phi \subseteq \Psi \in \mathbf{CON} \quad \Rightarrow \quad \Psi = \Phi.$$

The central pillar which supports the completeness proof is the following existence result.

1.4 LEMMA. (Basic Existence Result) *For each* $\Phi \in \textbf{CON}$ *there is some* $s \in \textbf{CON}$ *with* $\Phi \subseteq s$.

Proof. Let $\{\phi_r \mid r < \omega\}$ be an enumeration of all formulas. Let $(\Delta_r \mid r < \omega)$ be the ascending sequence of sets of formulas defined recursively by

$$\Delta_0 = \Phi$$

$$\Delta_{r+1} = \begin{cases} \Delta_r \cup \{\phi_r\} & \text{if this is in } \textbf{CON} \\ \Delta_r & \text{otherwise.} \end{cases}$$

Clearly $\Delta_r \in \textbf{CON}$ for all $r < \omega$, and hence

$$s = \bigcup \{\Delta_r \mid r < \omega\} \in \textbf{CON}.$$

Finally the construction ensures that $s \in \textbf{S}$. ∎

For any $s \in \textbf{S}$ let σ be the valuation given by

$$\sigma(P) = \begin{cases} \text{TRUE} & \text{if } P \in s \\ \text{FALSE} & \text{if } P \notin s \end{cases}$$

(for $P \in Var$). A routine induction now shows that σ is a model of (all the formulas in) s. This makes use of the fact that for $\Phi \in \textbf{S}$ the implications of the preservation properties are equivalences. Thus we have the following.

1.5 THEOREM. *Each* $\Phi \in \textbf{CON}$ *has a model.*

Finally we can achieve the desired completeness result.

1.6 THEOREM. (Completeness) *For each set of formulas Φ and formula ϕ the equivalence*

$$\Phi \vdash \phi \quad \Leftrightarrow \quad \Phi \models \phi$$

holds.

Proof. The implication ($\Rightarrow$) is soundness, so it suffices to prove ($\Leftarrow$). Thus suppose $\Phi \models \phi$. Then $\Phi \cup \{\neg\phi\}$ has no model and hence Theorem 1.5 gives

$$\Phi \cup \{\neg\phi\} \notin \textbf{CON}.$$

Thus

$$\Phi, \neg\phi \vdash \bot$$

and hence the Deduction Property gives

$$\Phi \vdash (\neg\phi \rightarrow \bot)$$

which (with an appropriate axiom) gives

$$\Phi \vdash (\neg\bot \rightarrow \phi).$$

Finally, since $\Phi \vdash \neg\bot$, we have $\Phi \vdash \phi$, as required. ∎

1.6   Exercises

1.1   Constructing formal derivations can be quite tricky.

(a) Using only the logical axioms (k, s), exhibit witnessing deductions for each of the following.

  (i) $\vdash \phi \to \phi$

  (ii) $\vdash (\psi \to \phi) . \to . (\theta \to \psi) \to (\theta \to \phi)$

  (iii) $\vdash \theta \to (\psi \to \phi) . \to . \psi \to (\theta \to \phi)$

  (iv) $\vdash (\theta \to \psi) . \to . (\psi \to \phi) \to (\theta \to \phi)$

  (v) $\vdash (\theta \to (\theta \to \psi)) \to (\theta \to \psi)$

What are the lengths of these various deductions?

(b) Use the Deduction Property to verify (i - v).

1.2   A set Φ of formulas is said to be *finitely satisfiable* if each finite subset of Φ has a model. Let CON be the set of all finitely satisfiable sets of formulas. Show that CON has the closure properties of Section 1.5, and hence prove the compactness theorem, namely that each finitely satisfiable set of formulas is satisfiable.

1.3   Let P, Q, and R be three finite, pairwise disjoint sets of variables. Let ϕ be a formula built up from $P \cup Q$, and let ψ be a formula built up from $Q \cup R$. Suppose that

$$\phi \to \psi$$

is a tautology.

Let Π and Σ be, respectively, the sets of all assignments

$$P \longrightarrow 2 \quad , \quad R \longrightarrow 2$$

where 2 is the truth object. Note that Π and Σ are finite. For each $\pi \in \Pi$ and $\sigma \in \Sigma$ let

$$\phi^\pi \quad , \quad \psi^\sigma$$

be the result of replacing each $P \in P$ by $\pi(P)$ and each $R \in R$ by $\sigma(R)$. Let

$$\lambda = \bigvee \{\phi^\pi \mid \pi \in \Pi\} \quad , \quad \rho = \bigwedge \{\psi^\sigma \mid \sigma \in \Sigma\}$$

(so that λ and ρ depend only on Q).

(a) Show that

$$\phi \to \lambda \quad , \quad \lambda \to \rho \quad , \quad \rho \to \psi$$

are tautologies.

(b) Show that for each formula θ built up from $\mathbf{Q}$, if both

$$\phi \to \theta \quad , \quad \theta \to \psi$$

are tautologies, then

$$\lambda \to \theta \quad , \quad \theta \to \rho$$

are also tautologies.

These provide an interpolation result for propositional logic.

Chapter 2

The modal language

2.1 Introduction

The propositional modal language is an extension of the pure propositional language formed by adding a battery of new 1-ary connectives (known informally as box connectives). Originally there was just one new connective $\square$, however for many purposes it is necessary to add several (possibly infinitely many) such connectives $[i]$, one for each element i of an index set I. Thus there are many possible modal languages, one for each index set I. The syntax, semantics, and proof systems associated with modal languages are designed to subsume those of the proposition language, in fact, propositional logic can be regarded as the extreme version of modal logic where $I = \emptyset$.

The element i of I are called *labels* and I itself is called the *signature* of the modal language. Thus two languages are identical precisely when they have the same signature. (We are never going to consider how one language may be be translated into another, so we need not worry about comparison of signatures.)

Unlike the propositional connectives $\neg$, $\rightarrow$, $\wedge$, $\vee$, $\top$, and $\bot$, the box connectives $[i]$ do not have a fixed interpretation. For each formula ϕ (of the modal language) we may use $[i]$ to obtain a new formula

$$[i]\phi \tag{2.1}$$

This may be read in several ways, and different readings suggest different semantics and proof systems.

In the original work on modal logic, there was just one box connective (i.e. I was a singleton), and the compound formula (2.1) was read variously as

ϕ is necessary,
ϕ is obligatory,
ϕ is known,

and later other readings such as

ϕ is provable (in some formal system of arithmetic)

13

were considered. A binary version of modal logic (called *tense logic*) has just
two labels

$$+ \text{ (for future)} \qquad \text{and} \qquad - \text{ (for past)}.$$

We may then read

$$[+]\phi \qquad\qquad\qquad\qquad [-]\phi$$

as

$$\phi \text{ is and always will be} \qquad \phi \text{ is and always was}.$$

In these two readings, it is seen quite clearly that $[+]$ and $[-]$ are intended as
kinds of universal quantifiers over time.

   More recently modal languages have been used to analyse the behaviour of
computer programs and the state transitions of (finite) automata. It is in these
applications where many different labels may be required. (In some cases these
applications may also require a further enrichment of this modal language, but
we won't be dealing with these applications in this book.)

   Observe that for each of the above readings there is also a dual complement
reading. Thus in the monomodal cases we have

$$\begin{array}{ll} \phi & \text{is possible,} \\ \phi & \text{is permissible,} \\ \neg\phi & \text{is unknown,} \end{array}$$

or

$$\phi \qquad \text{is consistent (with some formal system of arithmetic)}$$

and in the bimodal (tense) case we have

$$\begin{array}{cc} \phi \text{ will be} & \phi \text{ was} \\ \text{(at least once in the future)} & \text{(at least once in the past)} \end{array}$$

To handle these readings let

$$\langle i\rangle\phi \qquad \text{abbreviate} \qquad \neg\,[i]\neg\phi.$$

We may then check that each reading of $[i]\phi$ produces (via the dual comple-
ment) a reading of $\langle i\rangle\phi$. All the required properties of $\langle i\rangle\phi$ (which we read
informally as 'diamond ϕ') can be deduced from those of $[i]\phi$. It is also possible to introduce $\langle i\rangle$ as a new atomic primitive. There are various arguments
for and against these two approaches (abbreviation .v. primitive), and the
differences are most acute when dealing with proof systems. However, these
hardly matter in this book, so I have taken the approach which involves the
minimum amount of work, namely to treat $\langle i\rangle$ as an abbreviation.

   As suggested by the above readings, there is a strong analogy between

$$\text{the connective } [i] \qquad \text{and} \qquad \text{the quantifier } \forall$$

and between

$$\text{the connective } \langle i \rangle \quad \text{and} \quad \text{the quantifier } \exists.$$

The analogy will be made precise in Chapter 4 (when we describe the semantics of $[i]$ and $\langle i \rangle$). However, you should always keep the analogy in mind; it will help you understand many things.

2.2 The language defined

Let I be some fixed signature (i.e. indexing set of labels). The primitive symbols of the modal language with this signature are:

- The elements P of a fixed countable set Var of *variables*;

- The *propositional connectives*

$$\top, \bot, \neg, \rightarrow, \wedge, \vee$$

  of arity 0, 0, 1, 2, 2, and 2, respectively;

- The *box connectives*
$$[i]$$
  one for each label $i \in I$;

together with the *punctuation symbols* (and). The formulas of this language are then constructed in the expected way.

2.1 DEFINITION. The formulas of the language with signature I are obtained recursively using the following clauses.

atomic Each variable $P \in Var$ and each constant $\top$ and $\bot$ is a formula.

propositional For all formulas θ, ψ, ϕ each of

$$\neg\phi \quad , \quad (\theta \rightarrow \psi) \quad , \quad (\theta \wedge \psi) \quad , \quad (\theta \vee \psi)$$

is a formula.

modal For each formula ϕ and label i,

$$[i]\phi$$

is a formula.

Let $Form$ be the set of all formulas. ∎

Note that (as with the propositional language) the precise size of Var has not been specified. The most natural case is to take Var to be countably infinite, however, for some purposes (such as decidability matters) we may want Var to be finite. There is even a use for the case $Var = \emptyset$ e.g. in the analysis of concurrency. (The case where Var is uncountably infinite has some interest.)

When displaying formulas we will attempt to make them easier to read by the use of various conventions (such as the judicious omission of brackets). Also as suggested above we let

$$\langle i \rangle \phi \qquad \text{abbreviate} \qquad \neg\, [i]\neg\phi.$$

2.3 Some particular formulas

Several particular formulas (or rather shapes of formulas) play a prominent role in modal logic. In this section we gather together the majority of these shapes.

First some shapes which are concerned with just one label. In such circumstances we usually omit the label and write

$$\Box\phi \text{ for } [i]\phi \quad \text{and} \quad \Diamond\phi \text{ for } \langle i \rangle\phi.$$

This will help prevent displayed formulas from becoming cluttered with unnecessary symbols.

The first batch of shapes have names most of which are used for historical reasons (but now have rather limited significance). Thus, for an arbitrary formula ϕ, let $D(\phi)$, $T(\phi)$, ... , $M(\phi)$ be as follows:

$$
\begin{array}{rcl}
D(\phi) & : & \Box\phi \rightarrow \Diamond\phi \\
T(\phi) & : & \Box\phi \rightarrow \phi \\
B(\phi) & : & \phi \rightarrow \Box\Diamond\phi \\
4(\phi) & : & \Box\phi \rightarrow \Box\Box\phi \\
5(\phi) & : & \Diamond\phi \rightarrow \Box\Diamond\phi \\
P(\phi) & : & \phi \rightarrow \Box\phi \\
Q(\phi) & : & \Diamond\phi \rightarrow \Box\phi \\
R(\phi) & : & \Box\Box\phi \rightarrow \Box\phi \\
G(\phi) & : & \Diamond\Box\phi \rightarrow \Box\Diamond\phi \\
L(\phi) & : & \Box T(\phi) \rightarrow \Box\phi \\
M(\phi) & : & \Box\Diamond\phi \rightarrow \Diamond\Box\phi
\end{array}
$$

If we use two or more labels then we get a greater variety of formulas. For instance, shapes D, B, 5 and G can be generalized as follows:

$$
\begin{array}{rcl}
[i]\phi & \rightarrow & [j]\phi \\
\phi & \rightarrow & [i]\langle j \rangle\phi \\
\langle i \rangle\phi & \rightarrow & [j]\langle k \rangle\phi \\
\langle i \rangle[j]\phi & \rightarrow & [k]\langle l \rangle\phi
\end{array}
$$

Various other shapes will be important at one time or another, especially when particular applications are under consideration.

2.4 Substitution

Substitution is an important aspect of formal languages which, superficially, is easy to understand, but which has many pitfalls (down which many an author has disappeared). Most problems occur when there are both free and bound variables around (such as in the predicate calculus or the λ-calculus) but even in free variable systems such as the ones considered in this book, it is not entirely straight forward.

Consider a formula ϕ built up from the variables $P_1, \ldots, P_n$, that is, the only variables occurring in ϕ are amongst the ones listed. Suppose also that $\pi_1, \ldots, \pi_n$ is a list of formulas matching the list of variables. It is intuitively clear what we mean by the formula obtained from ϕ by simultaneously replacing P_1 by π_1, P_2 by π_2, $\ldots$, P_n by π_n. The resulting formula we may write as

$$\phi[P_1 := \pi_1, P_2 := \pi_2, \ldots, P_n := \pi_n].$$

(There are also several other equally cumbersome and uninformative notations in common use.) For example, let P and Q be distinct variables and let ψ be

$$(P \to Q).$$

Then, for formulas λ and μ,

$$\psi[P := \lambda, Q := \mu] \quad \text{is} \quad (\lambda \to \mu).$$

Even such simple examples can cause trouble. For instance, what happens if λ and μ also contain P and Q, and how do we handle iterated substitutions? Before you continue you should make sure you know why the two formulas

$$\psi[P := (Q \to P), Q := P]$$
$$\psi[P := (Q \to P), Q := Q][P := P, Q := P]$$

are

$$((Q \to P) \to P) \quad \text{and} \quad ((P \to P) \to P)$$

respectively.

The great majority of textbooks leave the notion of substitutions as an 'intuitively obvious' operation (and many don't consider it at all). Unfortunately there are one or two places where a more precise knowledge is required, and for these the only safe way is to give a formal definition. (This, of course, requires also that the definition is correct.)

There are several slightly different ways of handling substitution. The one described below is chosen because of its similarity with the application of a valuation (which we look at in Chapter 4).

2.2 DEFINITION. A *substitution* is a function

$$\sigma : Var \longrightarrow Form$$

Let Sub be the set of all such substitutions. ∎

   Thus a particular substitution σ assigns to each variable P a formula $\sigma(P)$.
We may then use σ to modify any formula ϕ by replacing each variable P
occurring in ϕ by $\sigma(P)$. This replacement must be done simultaneously for all
variables. Let us write ϕ^σ for the result of applying the substitution σ to ϕ.
For example, if

$$\sigma(P) = \lambda \quad , \quad \sigma(Q) = \mu$$

then

$$(P \to Q)^\sigma \quad , \quad (\lambda \to \mu).$$

This construction produces an operation

$$Form \times \mathsf{Sub} \longrightarrow Form$$
$$\phi \quad , \quad \sigma \longmapsto \phi^\sigma.$$

with the following formal definition.

2.3 DEFINITION. Let $\sigma \in \mathsf{Sub}$. For each formula ϕ the substitution instance
ϕ^σ is defined by recursion on ϕ using the following clauses.

(Const) For the constants

$$\top^\sigma = \top \quad , \quad \bot^\sigma = \bot.$$

(Var) For each variable P

$$\sigma(P) \quad , \quad \sigma(P).$$

(¬) For each formula $\phi = \neg\theta$

$$(\neg\phi)^\sigma = \neg\phi^\sigma.$$

($\wedge, \vee, \to$) For each formula θ, ψ

$$\begin{aligned}
(\theta \wedge \psi)^\sigma &= (\theta^\sigma \wedge \psi^\sigma) \\
(\theta \vee \psi)^\sigma &= (\theta^\sigma \vee \psi^\sigma) \\
(\theta \to \psi)^\sigma &= (\theta^\sigma \to \psi^\sigma).
\end{aligned}$$

(i) For each label i and formula ϕ

$$([i]\phi)^\sigma = [i]\phi^\sigma.$$

   where $[i]$ is the appropriate box connective. ∎

2.5 Two remarks

Every modal language has a modal-free part consisting of all the formulas which may be constructed without the use of a box (or diamond) symbol. These formulas are precisely the formulas of the (modal-free) propositional language of Chapter 1. Amongst these modal-free formulas we find the tautologies (i.e. those formulas which are true under all 2-valuations).

Given a modal-free formula ϕ and a substitution σ, we may apply σ to ϕ to obtain a modal formula ϕ^σ which, of course, may no longer be modal-free. When ϕ is a tautology we say ϕ^σ is *an instance of a tautology* or sometimes, rather loosely, simply a tautology. When we come to deal with the semantics of modal formulas (in Chapter 4) we will see that such generalized tautologies hold in all modal semantic situations.

Since formulas are constructed recursively on their complexity, many proofs about them proceed by induction on this complexity. For some of these proofs the notion of the *set $\Gamma(\phi)$ of subformulas of ϕ* is useful. This set is defined by:

(Const) For the constants

$$\Gamma(\top) = \{\top\} \quad , \quad \Gamma(\bot) = \{\bot\}.$$

(Var) For each variable P

$$\Gamma(P) = \{P\}.$$

($\neg$) For each formula $\phi = \neg\theta$

$$\Gamma(\phi) = \Gamma(\theta) \cup \{\phi\}.$$

($\wedge, \vee, \rightarrow$) For each formula $\phi = \theta * \psi$ where $*$ is a binary connective

$$\Gamma(\phi) = \Gamma(\theta) \cup \Gamma(\psi) \cup \{\phi\}.$$

(i) For each label i and formula $\phi = [i]\theta$

$$\Gamma(\phi) = \Gamma(\theta) \cup \{\phi\}.$$

In particular, note that $\phi \in \Gamma(\phi)$ for all ϕ.

2.6 Exercises

2.1 Many properties of formulas ϕ are defined by recursion on the complexity of ϕ, and then proof about these properties are achieved by induction on this complexity. For instance, with each formula ϕ we may associate three sets

$$Var(\phi) \quad , \quad Pos(\phi) \quad , \quad Neg(\phi)$$

of variables using the following clauses.

ϕ	$Var(\phi)$	$Pos(\phi)$	$Neg(\phi)$
$\bot, \top$	$\emptyset$	$\emptyset$	$\emptyset$
P	$\{P\}$	$\{P\}$	$\emptyset$
$\neg\theta$	$Var(\theta)$	$Neg(\theta)$	$Pos(\theta)$
$\theta \wedge \psi$	$Var(\theta) \cup Var(\psi)$	$Pos(\theta) \cup Pos(\psi)$	$Neg(\theta) \cup Neg(\psi)$
$\theta \vee \psi$	$Var(\theta) \cup Var(\psi)$	$Pos(\theta) \cup Pos(\psi)$	$Neg(\theta) \cup Neg(\psi)$
$\theta \rightarrow \psi$	$Var(\theta) \cup Var(\psi)$	$Neg(\theta) \cup Pos(\psi)$	$Pos(\theta) \cup Neg(\psi)$
$[i]\theta$	$Var(\theta)$	$Pos(\theta)$	$Neg(\theta)$

(a) Show that for each formula ϕ we have

$$Var(\phi) \;=\; Pos(\phi) \cup Neg(\phi)$$

and find an example with $Pos(\phi) \cap Neg(\phi) \neq \emptyset$.

(b) Give recursive definitions of the two set

$$Pos^+(\phi) \;=\; Var - Neg(\phi) \quad , \quad Neg^+(\phi) \;=\; Var - Pos(\phi)$$

where Var is the set of all variables.

2.2 For a given variable P, consider the four formulas

$$\phi_1 := P \quad , \quad \phi_2 := \neg P \quad , \quad \phi_3 := \neg P \rightarrow P \quad , \quad \phi_4 := P \rightarrow \neg P.$$

Compute $\phi_i[P := \phi_j]$ for all $1 \le i, j \le 4$.

2.3 For distinct variables P and Q let

$$\theta := P \quad , \quad \psi := Q \quad , \quad \phi := P \rightarrow Q.$$

For arbitrary formulas ρ and σ compute

(a) $\phi[P := \psi, Q := \theta][Q := \rho, P := \sigma]$

(b) $\phi[P := \psi[Q := \rho, P := \sigma], Q := \theta[Q := \rho, P := \sigma]]$

showing that the resulting formulas are the same.

2.4 Let σ and τ be a pair of substitutions which agree on the variables occurring in a given formula ϕ. Show that $\phi^\sigma = \phi^\tau$.

2.5 For substitutions σ and τ let $\tau \bullet \sigma$ be the substitution given by

$$(\tau \bullet \sigma)(P) \;=\; (\sigma(P))^\tau$$

for all variables P.

(a) Show that
$$(\phi^\sigma)^\tau \;=\; \phi^{\tau \bullet \sigma}$$
holds for all formulas ϕ.

(b) Show that
$$(\tau \bullet \sigma) \bullet \rho \;=\; \tau \bullet (\sigma \bullet \rho)$$
hold for all substitutions ρ, σ, τ.

2.6 Make sure you understand the notion of 'subformula'.

(a) For a variable P, write out $\Gamma(\mathrm{L}(P))$.

(b) Show that
$$\psi \in \Gamma(\phi) \;\Rightarrow\; \Gamma(\psi) \subseteq \Gamma(\phi)$$
(for arbitrary formulas ψ and ϕ).

Part II

Transition structures and semantics

This part introduces and develops the Kripke style semantics for modal languages.

The supporting structures for this semantics, here called *labelled transition systems* but sometimes called *frames* of *Kripke structures*, are described in Chapter 3. Then in Chapter 4 these structures are enriched by *valuations* which enable us to give the semantics of the languge (relative to an arbitrary valued structure). The semantics is given in terms of what is sometimes called a *forcing relation*. The concepts introduced in these two chapters are the most important in the whole book.

This Kripke (or forcing) semantics provides a link between the structures and the language, and we find that many property of thses structures can be captured by appropriate modal formulas. This idea, which is known as *correspondence theory*, is introduced and exemplified in Chapter 5. Chapter 6 is devoted to the proof of a correspondence result which covers many, but not all, of the cases.

Chapter 3

Labelled transition structures

3.1 Introduction

The central concept of this book is that of a *labelled transition structure* or, for brevity, simply a *structure*. These are the relational structures used to support the standard semantics of polymodal languages. In their monomodal form they are known as *Kripke structures* or *frames* (because of their supporting role). The introduction of these structures into modal logic around 1960 brought about a considerable amount of clarification and insight, and stimulated a rapid development of the subject. The use of these structures is a powerful tool and elevates the subject above the rather tedious symbol shuffling and philosophical ramblings that used to be its forte.

However, these structures are not just the tools of modal logic. They occur naturally in many parts of mathematics and computer science. For instance, partial orderings, equivalence relations, graphs, automata, and process algebras all give examples of these transition structures. This points to a true perspective of the relationship between modal languages and their semantic structures.

The objective of modal logic is not an analysis of modal languages; it is not the study of certain formal systems and the relationships between these; it is not the construction of different proof styles and rules of inference, etc: these are merely techniques developed to help the practitioner towards his central aim. The main objective of modal logic is, no more and no less, the study of labelled transition structures. The modal language and all its attachments is there simply to help out.

3.1 DEFINITION. Let I be a non-empty set. A *labelled transition structure* of signature I is a relational structure

$$\mathcal{A} = (A, \mathsf{A})$$

where A is a non-empty set and

$$\mathsf{A} = (\xrightarrow{i} \mid i \in I)$$

25

is an I-indexed family of binary relations $\overset{i}{\longrightarrow}$ on A. ∎

We will more often than not refer to a labelled transition structure as simply as a *structure*; this will save a bit of time and space.

For elements a and b of the structure $\mathcal{A}$ (i.e. members of the carrying set A) we write

$$a \overset{i}{\longrightarrow} b$$

to indicate that a and b stand in the relation $\overset{i}{\longrightarrow}$. In certain situations this notation can be rather cumbersome, so we also write

$$b \prec_i a$$

to convey precisely the same information. In particular, you should note how the position of the two elements have been interchanged; we will later make good use of this interchange.

In situations where we are concerned with just one of the distinguished relations, we will drop the affixed label i and write simply

$$a \longrightarrow b \quad \text{or} \quad b \prec a. \tag{3.1}$$

In particular we use this streamlined notation in the monomodal case, i.e. the case where the signature I is a singleton. In this case structures are sometimes called *frames*, the elements are called *possible worlds*, and the distinguish relation is known as the *accessibility relation*. When (3.1) holds we say world b is accessible from world a, or that a can see b.

It is often useful to think of a labelled transition structure (of a general signature) as a description of the possible states, or configurations, of a machine. Each element corresponds to a state of the machine, and each label corresponds to a possible external influence on the machine. We then read

$$a \overset{i}{\longrightarrow} b$$

as:

> When the machine is in a state a, a possible result of external influence i is a change to state b.

Note that the influence i need not have a unique effect on state a. There may be distinct states b and c with

$$a \overset{i}{\longrightarrow} b \quad , \quad a \overset{i}{\longrightarrow} c$$

in which case influence i will move the machine from state a to state b or state c (or possibly other states) in an indeterminate fashion. When this is not the case we have a deterministic machine.

3.2 DEFINITION. The structure $\mathcal{A}$ is *i-deterministic* (for a label i) if for all elements a, x, y

$$\left. \begin{array}{c} a \xrightarrow{\,i\,} x \\ a \xrightarrow{\,i\,} y \end{array} \right\} \quad \Rightarrow \quad x = y$$

The structure is *deterministic* if it is *i*-deterministic for all labels *i*. ∎

Observe that even when $\mathcal{A}$ is deterministic, for a given element a and label i, there need not be an element b with $b \prec_i a$. There can, however, be no more than one such element.

3.2 Some examples

Let us look at some simple examples of labelled transition structures.

Consider first the possible structures of signature I on the 1-element set $\{*\}$. For each label i we need to know whether or not

$$* \xrightarrow{\,i\,} *.$$

Thus we see that these structures are in bijective correspondence with the subsets of I.

Next consider the possible structures on a 2-element set. For each ordered pair a, b of elements and each label i we need to know whether or not

$$a \xrightarrow{\,i\,} b.$$

This can generate quite a lot of structures. For instance, with just two labels there are 2^8 such structures. However, not all of these are isomorphically distinct, for swapping the two elements may produce a structure that is essentially the same as another. You may like to list all of these structures and pick out the ones that are essentially different up to isomorphism, (then again, you may not).

Suppose there is just one label and consider the 5-element structure

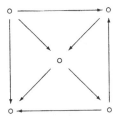

where no element is a reflexive point. Notice that there are infinite paths going through this structure, all of these pass through the bottom right element infinitely often, may begin at the top left element otherwise will not pass through that element, and will not pass through the bottom left element.

The set N of natural numbers carries many different transition relations. For instance, we may let

$$x \longrightarrow y$$

mean any of

$$x < y \, , \, x \leq y \, , \, x > y \, , \, x \geq y$$

$$y = x + 1 \, , \, y \leq x + 1 \, , \, |\, x - y \,| \leq 2$$

to obtain seven such relations. All of these will be used later to illustrate various facets of modal logic.

In the next section we will see how a labelled transition structure can be converted into a different, but equivalent, kind of structure called a *modal algebra*. For many purposes these algebras are easier to deal with but are not absolutely necessary. Modal algebras are introduced now because, logically, this is where they should be. However, the benefits of them are not needed until quite a bit later so, for pedagogical purposes you may want to skip the rest of this chapter and go straight to Chapter 4 . Of course, this will mean you have to return to this chapter at a later time.

3.3   Modal algebras

There is a different way of looking at structures (transition structures) which is sometimes more convenient. We will give a construction which converts each structure into an enriched boolean algebra which for many applications is easier to work with.

Thus, fix a structure $\mathcal{A}$ (of signature I) and consider the power set $\mathcal{P}A$ of the carrying set A of $\mathcal{A}$. The elements of $\mathcal{P}A$ are all the subsets of A (including the empty set $\emptyset$ and A itself). These subsets $X, Y, Z, \ldots$ are partially ordered by inclusion

$$X \subseteq Y$$

and can be combined by union, intersection, and complementation

$$X \cup Y \quad , \quad X \cap Y \quad , \quad \neg X = A - X$$

to form new subsets. Thus we may view $\mathcal{P}A$ as a boolean algebra.

We enrich this algebra $\mathcal{P}A$ by a family of operations $[i]$, one for each label i. The operation $[i]$ is obtained in a predetermined fashion from the relation $\overset{i}{\longrightarrow}$. The resulting algebra

$$(\mathcal{P}A, (\, [i] \mid i \in I))$$

is called the *modal algebra* induced by $\mathcal{A}$.

For each subset $X \subseteq A$, the set $[i]X$ consists precisely of those elements $a \in A$ such that every elements x of A with $a \xrightarrow{i} x$ is automatically in X. If we use the alternative notation '$\prec_i$' the definition can be given succinctly as

$$a \in [i]X \Leftrightarrow (\forall x \prec_i a)[x \in X] \tag{3.2}$$

Here the right hand side is an abbreviated form of

$$(\forall x)[x \prec_i a \Rightarrow x \in X]$$

where the quantified variable x ranges over A. Let us look at some simple properties of this operation.

3.3 PROPOSITION. *The operation $[i]$ satisfies*

$$[i]A = A,$$

is monotone, and satisfies

$$[i](X \cap Y) = [i]X \cap [i]Y$$

for all $X, Y \in \mathcal{P}A$.

Proof. The first identity is trivial. For the monotone property, if $X \subseteq Y$ then, for each $a \in A$ we have

$$
\begin{aligned}
a \in [i]X &\Rightarrow (\forall x \prec_i a)[x \in X] \\
&\Rightarrow (\forall x \prec_i a)[x \in Y] \quad \Rightarrow \quad a \in [i]Y
\end{aligned}
$$

so that $[i]X \subseteq [i]Y$.

For the second identity observe that, for arbitrary $X, Y \in \mathcal{P}A$, monotonicity gives

$$[i](X \cap Y) \subseteq [i]X \cap [i]Y.$$

Conversely, for arbitrary $a \in A$,

$$a \in [i]X \cap [i]Y \Rightarrow a \in [i]X \text{ and } a \in [i]Y$$

$$
\Rightarrow
\begin{cases}
(\forall x \prec_i a)[x \in X] \\
\text{and} \\
(\forall x \prec_i a)[x \in Y]
\end{cases}
$$

$$\Rightarrow (\forall x \prec_i a)[x \in X \text{ and } x \in Y]$$

$$\Rightarrow (\forall x \prec_i a)[x \in X \cap Y] \qquad \Rightarrow \quad a \in [i](X \cap Y)$$

as required. ∎

In general the operation $[i]$ has no other properties beyond those suggested by Proposition 3.3 unless the parent relation $\xrightarrow{i}$ is restricted in some way. For instance consider the 2-element set

$$A = \{u, v\}$$

furnished with the relation $\prec$ where

$$u \prec v \prec v$$

but no other pairs satisfy $\prec$. Then $\mathcal{P}A$ has boolean structure

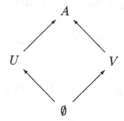

where

$$U = \{u\} \quad , \quad V = \{v\}$$

and we easily check that

$$\square\emptyset = \square U = \square V = U.$$

In particular we see that

$$\square\emptyset \neq \emptyset$$

and

$$\square(U \cup V) \neq \square U \cup \square V.$$

3.4   Various correspondences

For an arbitrary structure $\mathcal{A}$, some properties of the operation $\square$ correspond precisely to those of the parent relation $\prec$. We will describe some of these, but to do this we need some terminology.

We assume known the properties of *reflexivity* and *transitivity* of a relation. Also, we say the relation $\prec$ is *pathetic* if

$$x \prec a \quad \Rightarrow \quad x = a$$

(for all $x, a \in A$); and we say $\prec$ is *dense* if for each $a, b \in A$ with $b \prec a$, there is some $x \in A$ with $b \prec x \prec a$. Note that every reflexive relation is dense, and every pathetic relation is transitive. We say the operation $\square$ is *deflationary* or *inflationary* if, for each $X \in \mathcal{P}A$ we have, respectively,

$$\square X \subseteq X \quad , \quad X \subseteq \square X.$$

We say $\square$ is *nearly deflationary* or *nearly inflationary* if, for each $X \subseteq A$ we have, respectively

$$\square\,\square X \subseteq \square X \quad , \quad \square X \subseteq \square\,\square X.$$

Clearly, every deflationary operation is nearly deflationary and every inflationary operation is nearly inflationary.

At several places we will need to consider iterated applications of operations, and for this some abbreviations are useful. We write

$$\square^2 X \quad \text{for} \quad \square\,\square X$$

so the two nearly properties become

$$\square^2 X \subseteq \square X \quad , \quad \square X \subseteq \square^2 X.$$

How do these various properties relate to each other? They come in equivalence pairs.

3.4 PROPOSITION. *In the list*

$\prec$	$\square$
Reflexive	*Deflationary*
Transitive	*Nearly inflationary*
Pathetic	*Inflationary*
Dense	*Nearly deflationary*

the horizontal pairs of properties correspond, that is, $\prec$ has a property precisely when $\square$ has the paired property.

Proof. Suppose first that $\prec$ is reflexive and consider any situation

$$a \in \square X$$

(where $a \in A$ and $X \subseteq A$). Then, for all $x \in A$,

$$x \prec a \;\Rightarrow\; x \in X.$$

But (since $\prec$ is reflexive) $x = a$ satisfies the hypothesis of this implication, so $x = a$ satisfies the conclusion. Thus $a \in X$. This shows that $\square X \subseteq X$, i.e. that $\square$ is deflationary.

Conversely, suppose that $\square$ is deflationary and, for a given $a \in A$, set

$$X = \{x \in A \mid x \prec a\}.$$

Then, by construction, $a \in \Box X$, so that (since $\Box X \subseteq X$) we have $a \in X$ and hence $a \prec a$. Thus $\prec$ is reflexive.

The other three equivalences are verified in a similar fashion.

For instance, suppose that $\prec$ is dense and consider any

$$a \in \Box^2 X$$

(for $a \in A$ and $X \subseteq A$). To show that $a \in \Box X$ (and hence that $\Box$ is nearly deflationary) consider any $x \prec a$. We require $x \in X$. But $\prec$ is dense, so there is some $y \in A$ with $x \prec y \prec a$. Then, with $Y = \Box X$, we have $a \in \Box Y$, so that $y \in Y = \Box X$, and hence $x \in X$, as required.

Conversely, suppose that $\Box$ is nearly dense and, for a given $a \in A$, let Y be the subset of A given by

$$y \in Y \quad \Leftrightarrow \quad (\exists x)[y \prec x \prec a].$$

Then, by construction $a \in \Box^2 Y$ and hence (since $\Box$ is nearly deflationary) $a \in \Box Y$, so that (by construction of $\Box$)

$$b \prec a \Rightarrow b \in Y \Rightarrow (\exists x)[b \prec x \prec a]$$

which verifies the required density property. ■

It is worth remembering here that a topological interior operation on $\mathcal{P}A$ is an operation $\Box$ which is deflationary and idempotent and satisfies the two properties of Proposition 3.3. In the presence of the deflationary property, idempotency is equivalent to being nearly inflationary. Thus topological spaces arise from monomodal structures

$$\mathcal{A} = (A, \longrightarrow)$$

where $\longrightarrow$ is a reflexive and transitive relation (i.e. a pre-order). This observation indicates how modal logic has a much greater depth than modal-free propositional logic.

3.5 The diamond operation

Take another look at the equivalence (3.2) defining $[i]$ from $\prec_i$. Notice there is a universal quantifier on the right hand side. What would happen if we changed this to an existential quantifier? Thus suppose we define the operation $\langle i \rangle$ by

$$a \in \langle i \rangle X \quad \Leftrightarrow \quad (\exists x \prec_i a)[x \in X]$$

(for $a \in A$ and $X \subseteq A$). How are $\langle i \rangle$ and $[i]$ related? The answer should be obvious to you.

Recall that $\mathcal{P}A$ carries a complementation operation. Let us denote this by '$\neg$' so that, for $a \in A$ and $X \subseteq A$,

$$a \in \neg X \quad \Leftrightarrow \quad a \notin X.$$

This operation is an involution, i.e.

$$\neg\neg X = X$$

(for all $X \in \mathcal{P}A$).

3.5 PROPOSITION. *The two operations*

$$[i] \quad , \quad \langle i \rangle$$

are dual complements, that is

$$\neg\, [i] X \;=\; \langle i \rangle \neg X$$

holds for all $X \subseteq A$.

Proof. We may omit the label i (since there is no other one involved). It suffices to recall how negation and quantifiers interact. In particular the three expressions

$$\neg(\forall x)[x \prec a \Rightarrow x \in X]$$
$$(\exists x)\neg[x \prec a \Rightarrow x \in X]$$
$$(\exists x)[x \prec a \wedge x \notin X]$$

are logically equivalent. Thus, for $a \in A$ and $X \subseteq A$, we have

$$a \in \neg\, \square X \quad \Leftrightarrow \quad \neg(\forall x \prec a)[x \in X]$$
$$\Leftrightarrow \quad (\exists x \prec a)[x \in \neg X] \quad \Leftrightarrow \quad a \in \Diamond \neg X$$

which is the required result. ∎

Notice how the symbol '$\neg$' has been used in two different ways in this proof. On the one hand it has been used for the complementation operation on $\mathcal{P}A$ and, on the other, for the informal logical connective 'not'. It is precisely because these two notions match exactly that enables us to overload the symbol '$\neg$', and so expose the reason behind Proposition 3.5.

Some properties of the relation $\prec$ are best expressed using the operation $\Diamond$ rather than $\square$, and some properties require both. For instance, let us look at determinism (as introduced in Definition 3.2).

3.6 PROPOSITION. *The relation $\prec$ is deterministic precisely when*

$$\Diamond X \subseteq \square X$$

holds for all $X \subseteq A$.

Proof. Suppose first that $\prec$ is deterministic and situation consider any

$$a \in \Diamond X$$

(for some $a \in A$ and $X \subseteq A$). Then, by definition of $\Diamond$, there is some $b \in A$ with

$$b \prec a \quad , \quad b \in X.$$

But then, for each $x \in A$, since $\prec$ is deterministic,

$$x \prec a \Rightarrow x = b \Rightarrow x \in X$$

so that $a \in \Box X$, as required.

Conversely, consider any pair of elements a, b with

$$b \prec a.$$

Then, with $X = \{b\}$, we have

$$a \in \Diamond X \subseteq \Box X$$

(where here the local hypothesis has been invoked). This gives

$$x \prec a \Rightarrow x \in X \Rightarrow x = b$$

(for $x \in A$), and hence $\prec$ is deterministic. ∎

In due course we will see many other pairs of corresponding properties.

3.6   The structure regained

We have seen how every transition structure produces a modal algebra (of the same signature). This passage from structure to algebra does not lose any information, as the following result shows.

3.7 LEMMA. *For all $a, b \in A$*

$$b \prec a \Leftrightarrow (\forall X)[a \in \Box X \Rightarrow b \in X]$$

(where X ranges over $\mathcal{P}A$). Thus the operation $\prec$ can be retrieved from the operation $\Box$ on $\mathcal{P}A$.

Proof. The implication $\Rightarrow$ follows from the definition of $\Box$. Conversely, suppose the right hand condition holds and consider the set

$$X = \{x \in A \mid x \prec a\}.$$

Then, by construction, we have $a \in \bigcap X$, and hence $b \in X$, to give $b \prec a$, as required. ∎

As a final remark, observe how the introduction of these modal algebras opens up the possibility of a more general approach. The power set $\mathcal{P}A$ is a particular kind of boolean algebra, so we could, if required, consider general boolean algebras enriched by suitable operations. These more general modal algebras become necessary in later, more delicate, parts of modal logic; but they are barely required in this book.

3.7 Exercises

3.1 Let $A = \{u, v\}$ where u and v are distinct elements and consider the power set algebra $\mathcal{P}A$.

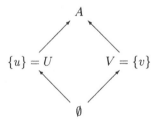

The set A carries 16 different transition relations corresponding to 16 different operations on $\mathcal{P}A$. These are listed in the following table. In the list of transition structures we use the convention

∘ = non-reflexive element , • = reflexive element.

The list of modal algebras gives the value of $\bigcap X$ below each set $X \in \{\emptyset, U, V\}$. Verify that this list is correct.

	Structure		Algebra		
	u	v	$\emptyset$	U	V
(1)	○	○	A	A	A
(2)	○	●	U	U	A
(3)	●	○	V	A	V
(4)	●	●	$\emptyset$	U	V
(5)	○ ←——— ○		U	A	U
(6)	○ ———→ ○		V	V	A
(7)	○ ←——— ●		U	U	U
(8)	○ ———→ ●		$\emptyset$	$\emptyset$	A
(9)	● ←——— ○		$\emptyset$	A	$\emptyset$
(10)	● ———→ ○		V	V	V
(11)	● ←——— ●		$\emptyset$	U	$\emptyset$
(12)	● ———→ ●		$\emptyset$	$\emptyset$	V
(13)	○ ←——→ ○		$\emptyset$	V	U
(14)	○ ←——→ ●		$\emptyset$	$\emptyset$	U
(15)	● ←——→ ○		$\emptyset$	V	$\emptyset$
(16)	● ←——→ ●		$\emptyset$	$\emptyset$	$\emptyset$

Note that the pairs (2,3), (5,6), (7,10), (8,9), (11,12), and (14,15) are isomorphic (via a swap of u and v). Thus the list contains only 10 essentially different transition structures.

3.2 Consider the set $A = \{u, v, w\}$ of three distinct elements. Show that A carries 256 different transition relations, although some of the resulting structures are isomorphic (via a permutation of A).

3.3 Let $\mathcal{A}$ be a monomodal transition structure with associated diamond operation $\Diamond$ on $\mathcal{P}A$. Show that $\Diamond$ is monotone and satisfies

$$\Diamond\emptyset = \emptyset \quad , \quad \Diamond(X \cup Y) = \Diamond X \cup \Diamond Y$$

for all $X, Y \in \mathcal{P}A$.

3.4 Let $\mathcal{A}$ be a monomodal structure whose associated transition relation is reflexive and symmetric.

(a) Show that

   (i) $\Box X \subseteq \Diamond \Box X \subseteq X \subseteq \Box \Diamond X \subseteq \Diamond X$

   (ii) $\Box \Diamond \Box X = \Box X$

for each $X \in \mathcal{P}A$.

(b) Show also that $\square$ need not be idempotent.

3.5 For an arbitrary set A let $[\bullet]$ be an operation on A which is monotone and satisfies

$$[\bullet]A = A \quad , \quad [\bullet](X \cap Y) = [\bullet]X \cap [\bullet]Y$$

for all $X, Y \in \mathcal{P}A$. Let $\longrightarrow$ be the transition relation on A given by

$$a \longrightarrow b \iff (\forall X \in \mathcal{P}A)[a \in [\bullet]X \Rightarrow b \in X]$$

and let $\square$ be the modal operation on $\mathcal{P}A$ induced by $\longrightarrow$.

(i) Show that $[\bullet]X \subseteq \square X$ holds for all $X \in \mathcal{P}A$.

(ii) Show that the two operations are the same precisely when

$$[\bullet]\bigcap \mathcal{X} = \bigcap\{ [\bullet]X \mid X \in \mathcal{X}\}$$

holds for all $\mathcal{X} \subseteq \mathcal{P}A$.

(iii) Give an example where $\square$ and $[\bullet]$ are not the same.

Chapter 4

Valuation and satisfaction

4.1 Introduction

For each signature I we have now defined two quite different entities; the poly-modal language of signature I, and the class of structures (labelled transition structures) of signature I. These two entities must now be made to interact. Thus the structures will be made to support a semantics for the language, or, equivalently, the language will be used to describe properties of structures.

The polymodal language has the usual propositional facilities together with a family of new 1-ary connectives $[i]$, one for each label i. We now wish to evaluate each formula of this language, i.e. determine whether or not a formula ϕ is TRUE or FALSE. Of course, this can not be done in isolation, we need to work in an appropriate context. To determine the truth value of ϕ we need three pieces of information together with an agreed procedure for using the information.

1. We need to know the truth values of the variables appearing in ϕ. As in the propositional case this information will be conveyed by a *valuation*, however, these modal valuations are more complicated than the propositional versions.

2. We need to know how to handle the propositional connectives. This will be done in exactly the same way as the propositional language (i.e. using the defining truth tables of the connectives). In this sense, modal logic subsumes propositional logic.

3. We need to know how to handle the modal connectives $[i]$. This will be done by working relative to a given structure $\mathcal{A}$. The relation $\xrightarrow{i}$ of this structure will control the connective $[i]$. The precise way this is done will be described in due course.

Thus, we are going to define a relation

$$(\mathcal{A}, \alpha, a) \Vdash \phi \tag{4.1}$$

39

between

structures $\mathcal{A}$,   valuations α ,   elements a of $\mathcal{A}$,   and formulas ϕ.

This relation may be read as:

> Under the circumstances determined by $(\mathcal{A}, \alpha, a)$, the formula ϕ is forced to be true.

As can be expected, this relation (4.1) is defined by recursion on the complexity of ϕ. In this recursion, the two parameters $\mathcal{A}$ and α are held fixed throughout, but the parameter a must be allowed to vary through all elements of $\mathcal{A}$.

In the majority of places where we use this satisfaction relation, we may suppress the parameters $\mathcal{A}$ and α and abbreviate (4.1) to

$$a \Vdash \phi.$$

We may also read this as

$$a \quad \text{forces} \quad \phi$$

this saves quite a bit of space (and mental breath) and allows us to concentrate on the important point, namely how the element a regards the truth status of ϕ. In circumstances where it seems helpful or will avoid misunderstandings, we will use the expanded form (4.1).

What is the appropriate notion of a model valuation of α? The information that α must provide is, for each variable P, at which elements of the supporting structure $\mathcal{A}$ is P regarded as TRUE, and at which elements is P regarded as FALSE. This information is supplied in the following fashion.

4.1 DEFINITION. A *valuation* α on a structure is an assignment

$$\alpha : Var \longrightarrow \mathcal{P}A$$

from variables P to subsets $\alpha(P)$ of A. The pair $(\mathcal{A}, \alpha)$ is then a *valued structure*. ∎

The idea here is that the variable P is regarded as TRUE at the element a precisely when $a \in \alpha(P)$. This provides the base case in the definition of (4.1), namely

$$a \Vdash P \quad \Leftrightarrow \quad a \in \alpha(P).$$

The recursion steps across the propositional connectives are the obvious ones, so all that remains (to complete the definition of (4.1)) is to describe how to handle

$$a \Vdash [i]\phi$$

for a label i and formula ϕ. Before we do this (in the next section) it is worthwhile looking at the nature of the structures involved.

For each signature I we have introduced three different *kinds* of associated structures. Firstly, we have the unadorned structures $\mathcal{A}$, i.e. the labelled transition structures described in Chapter 3. Each such structure can be enriched by a valuation α to form a *valued structure* $(\mathcal{A}, \alpha)$; and then we may distinguish a particular element a to form a *pointed valued structure* $(\mathcal{A}, \alpha, a)$. It is important not to confuse these three different kinds. We can not say that one kind is more important than the others; all three kinds have a role to play.

4.2 The basic satisfaction relation

So how do we handle the passage across a box $[i]$? To determine whether or not

$$a \Vdash [i]\phi$$

holds we must survey all the elements x with $a \overset{i}{\longrightarrow} x$ and for each such x determine whether or not

$$x \Vdash \phi$$

holds. Thus, the full and precise definition of $\Vdash$ is as follows.

4.2 DEFINITION. Let $(\mathcal{A}, \alpha)$ be a given valued structure. The relation

$$a \Vdash \phi$$

between elements a of $\mathcal{A}$ and formulas ϕ is defined by recursion on ϕ (with variations of the parameter a) using the following clauses.

(**Const**) For the constants

$$a \Vdash \top \quad , \quad \text{not}[a \Vdash \bot].$$

(**Var**) For each variable P

$$a \Vdash P \Leftrightarrow a \in \alpha(P).$$

($\neg$) For each formula ϕ

$$a \Vdash \neg\phi \Leftrightarrow \text{not}[a \Vdash \phi].$$

($\wedge, \vee, \rightarrow$) For all formulas θ, ψ

$$a \Vdash (\theta \wedge \psi) \Leftrightarrow a \Vdash \theta \text{ and } a \Vdash \psi$$
$$a \Vdash (\theta \vee \psi) \Leftrightarrow a \Vdash \theta \text{ or } a \Vdash \psi$$
$$a \Vdash (\theta \rightarrow \psi) \Leftrightarrow a \Vdash \psi \text{ whenever } a \Vdash \theta.$$

(i) For each label i and formula ϕ

$$a \Vdash [i]\phi \Leftrightarrow (\forall x \prec_i a)[x \Vdash \phi]$$

where the quantified variable x ranges over A (the carrying set of $\mathcal{A}$). ■

The right hand side of the equivalence in clause (i) is an abbreviated version of

$$(\forall x)[x \prec_i a \;\Rightarrow\; x \Vdash \phi].$$

The condensed version will prove to be very convenient in many computations.

In Section 4.1 we gave the formal reading of (4.1) together with a suggested shortening 'a forces ϕ'. For variety we will also read (4.1) as

$$
\begin{array}{rcl}
 & \phi & \text{is valid at } \; a, \\
 & \phi & \text{holds at } \; a, \\
(\mathcal{A}, \alpha, a) & \text{satisfies} & \phi, \\
(\mathcal{A}, \alpha, a) & \text{models} & \phi,
\end{array}
$$

or in various other similar ways.

Observe that each pointed valued structure $(\mathcal{A}, \alpha, a)$ gives us a 2-valuation ν_a where, for a variable P,

$$\nu_a(P) = \begin{cases} \text{TRUE} & \text{if } a \in \alpha(P) \\ \text{FALSE} & \text{if } a \notin \alpha(P). \end{cases}$$

We then see that for each propositional (i.e. box-free) formula ϕ,

$$a \Vdash \phi \quad \Leftrightarrow \quad [\![\phi]\!]_{\nu_a} = \text{TRUE}$$

where, of course, $[\![\cdot]\!]_{\nu_a}$ is the assignment induced by the 2-valuation ν_a as constructed in Chapter 1. In particular we obtain the following.

4.3 PROPOSITION. *If ϕ is a propositional tautology then ϕ is satisfied by every pointed valued structure.*

4.3   Some examples

The relation $\Vdash$ is probably the single most important notion in the whole of this book (and, in fact, the whole of modal logic). It is therefore worth spending some time looking at particular examples to help us develop a feel for the relation.

Consider the 4-element monomodal structure

where no element is reflexive. Consider also any valuation α with

$$\alpha(P) \quad = \quad \{a, c\}$$

for some variable P. Thus

$$a \Vdash P \quad , \quad b \Vdash \neg P \quad , \quad c \Vdash P \quad , \quad d \Vdash \neg P.$$

We also see that

$$a \Vdash \Box \neg P \quad , \quad b \Vdash \Box P \quad , \quad c \Vdash \Box P$$

for the only element x with, respectively

$$a \longrightarrow x \quad , \quad b \longrightarrow x \quad , \quad c \longrightarrow x$$

is

$$x = b \quad , \quad x = c \quad , \quad x = a.$$

Next we see that

$$a \Vdash \Box^2 P \quad , \quad b \Vdash \Box^2 P \quad , \quad c \Vdash \Box^2 \neg P.$$

For instance, to verify the first we must show that for all pairs x, y with

$$a \longrightarrow x \longrightarrow y$$

we have $y \Vdash P$. But the only such pair is $x = b$ and $y = c$, so we are done. A similar argument verifies the other two cases.

The element d is slightly more interesting. If we develop all the paths starting from d then we get

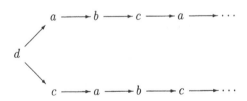

from which we find that

$$d \Vdash \Box P \quad , \quad d \Vdash \neg \Box^2 P \quad , \quad d \Vdash \neg \Box^3 P \quad , \quad d \Vdash \Box^4 P$$

etc.

Notice that the only path of length 3 starting from a is

$$a \longrightarrow b \longrightarrow c \longrightarrow a.$$

Thus, for any formula ϕ,

$$a \Vdash \Box \phi \quad \Leftrightarrow \quad b \Vdash \phi$$
$$a \Vdash \Box^2 \phi \quad \Leftrightarrow \quad c \Vdash \phi$$
$$a \Vdash \Box^3 \phi \quad \Leftrightarrow \quad a \Vdash \phi$$

so that

$$a \Vdash (\Box^3\phi \leftrightarrow \phi).$$

For similar reasons we see that $\Box^3\phi \leftrightarrow \phi$ also holds at b and c. Furthermore this argument is valid no matter which valuation is involved.

A similar, but slightly more complicated argument shows that

$$d \Vdash (\Box^4\phi \leftrightarrow \Box\phi)$$

and again this is independent of the valuation involved. This shows that the formula $\Box^4\phi \leftrightarrow \Box\phi$ holds at every element of $\mathcal{A}$ no matter which valuation is carried. In the notation introduced later (in the next section), this shows that

$$\mathcal{A} \Vdash^u \Box^4\phi \leftrightarrow \Box\phi.$$

As explained in Chapter 2, the diamond $\langle i \rangle$ may be introduced into the polymodal language as an abbreviation in the sense that, for each formula ϕ

$$\langle i \rangle \phi \quad \text{abbreviates} \quad \neg\,[i]\neg\phi.$$

It is instructive to see how the forcing relation $\Vdash$ handles this. The result shouldn't be a surprise.

4.4 PROPOSITION. *Let $(\mathcal{A}, \alpha)$ be a valued structure. Then the equivalence*

$$a \Vdash \langle i \rangle \phi \quad \Leftrightarrow \quad (\exists x \prec_i a)[x \Vdash \phi]$$

holds for all labels i, elements a, and formulas ϕ.

Proof. We use the standard manipulation of quantifiers. Using '$\neg$' for both the formal and the informal negation we have, applying the appropriate clauses of Definition 4.2

$$
\begin{aligned}
a \Vdash \langle i \rangle \phi &\Leftrightarrow a \Vdash \neg\,[i]\neg\phi \\
&\Leftrightarrow \neg[a \Vdash [i]\neg\phi] \\
&\Leftrightarrow \neg(\forall x \prec_i a)[x \Vdash \neg\phi] \\
&\Leftrightarrow (\exists x \prec_i a)\neg[x \Vdash \neg\phi] \Leftrightarrow (\exists x \prec_i a)[x \Vdash \phi]
\end{aligned}
$$

as required. ∎

It is worth comparing the equivalence of Proposition 4.4 with clause (i) of Definition 4.2. There are many similarities between the box connective $\Box$ and the universal quantifier $\forall$, and between the diamond connective $\Diamond$ and the existential quantifier $\exists$. Once this has been grasped, many of the computations of modal logic become almost routine. For instance, it is easy to check that

$$
\begin{aligned}
a \Vdash [i][j]\phi &\Leftrightarrow (\forall x \prec_i a)(\forall y \prec_j x)[y \Vdash \phi] \\
a \Vdash [i]\langle j \rangle \phi &\Leftrightarrow (\forall x \prec_i a)(\exists y \prec_j x)[y \Vdash \phi] \\
a \Vdash \langle i \rangle [j]\phi &\Leftrightarrow (\exists x \prec_i a)(\forall y \prec_j x)[y \Vdash \phi] \\
a \Vdash \langle i \rangle \langle j \rangle \phi &\Leftrightarrow (\exists x \prec_i a)(\exists y \prec_j x)[y \Vdash \phi].
\end{aligned}
$$

There are several particular valuations which ensure that some simple formulas are valid. For instance, for a given element a of the structure $\mathcal{A}$ and for a given variable P, consider any valuation α such that $\alpha(P) = \{a\}$. Then, for each $x \in A$,

$$x \Vdash P \quad \Leftrightarrow \quad x = a$$

and, trivially,

$$a \Vdash P.$$

Similarly, with a valuation α such that

$$\alpha(P) = \{x \in A \mid x \prec a\}$$

we have, for $x \in A$,

$$x \Vdash P \quad \Leftrightarrow \quad x \prec a$$

and

$$a \Vdash \square P.$$

In the same way, with a valuation such that

$$x \Vdash P \quad \Leftrightarrow \quad (\exists y)[x \prec y \prec a]$$

we have

$$a \Vdash \square^2 P.$$

A similar, but slightly more complicated, example deserves a little more formality. In this example we use the $*$-closure (i.e. reflexive, transitive closure) $\prec*$ of a relation $\prec$. If you are not familiar with this notion, a full discussion is given later in Chapter 14, Section 14.2.

4.5 PROPOSITION. *For a given element a of a structure $\mathcal{A}$ and a given variable P, consider any valuation on $\mathcal{A}$ such that*

$$x \Vdash P \quad \Leftrightarrow \quad (\forall y)[y \prec* x \Rightarrow y \prec a]$$

(for $x \in A$). Then

$$a \Vdash \square(\square P \to P)$$

holds.

Proof. Consider $b \prec a$ with

$$b \Vdash \square P.$$

We require that $b \Vdash P$. To this end consider any $y \prec* b$. Then, either

$$y = b \quad \text{or} \quad (\exists x)[y \prec* x \prec b]$$

so that either

$$y \prec a \quad \text{or} \quad (\exists x)[y \prec* x \Vdash P]$$

and hence, in both cases, $y \prec a$. Thus $b \Vdash P$, as required. ∎

4.4   The three satisfaction relations

So far we have defined only the basic satisfaction relation

$$(\mathcal{A}, \alpha, a) \Vdash \phi$$

for a pointed valued structure. We now allow this to ramify into three related satisfaction relations

$$\Vdash^p \quad , \quad \Vdash^v \quad , \quad \Vdash^u$$

all of which are useful at one place or another. It is extremely important that you learn to distinguish between these three satisfaction relations; the second and third are both derived from the first, but none can be said to be more fundamental than the other two.

The relation $\Vdash^p$ is just the relation $\Vdash$, and so it is used with pointed valued structures (hence the decoration p).

The relation $\Vdash^v$ is used with valued structures and is formed from $\Vdash^p$ by quantifying out the point. Thus, by definition,

$$(\mathcal{A}, \alpha) \Vdash^v \phi \quad \Leftrightarrow \quad (\forall a)[(\mathcal{A}, \alpha, a) \Vdash^p \phi]$$

where the quantified variable a ranges over all elements of $\mathcal{A}$.

Finally the relation $\Vdash^u$ is used with unadorned structures and is formed from $\Vdash^v$ by quantifying out the valuation. Thus

$$\mathcal{A} \Vdash^u \phi \quad \Leftrightarrow \quad (\forall \alpha)[(\mathcal{A}, \alpha) \Vdash^v \phi]$$

where the quantified variable α ranges over all valuations on $\mathcal{A}$.

All three give explications of the modelling process. Thus when we speak of 'a model' of a formula ϕ we could mean an unadorned structure, a valued structure, or a pointed valued structure. In all cases we should make sure we understand exactly which notion is intended.

Let us look at some examples of the differing properties of $\Vdash^p$, $\Vdash^v$, and $\Vdash^u$.

Consider the 4-element example $(\mathcal{A}, \alpha)$ described at the beginning of Section 4.3. There we have

$$(\mathcal{A}, \alpha, a) \Vdash^p P \quad \text{but} \quad \text{not}[(\mathcal{A}, \alpha) \Vdash^v P]$$

(for consider the element b). Notice also that

$$\text{not}[(\mathcal{A}, \alpha) \Vdash^v \neg P]$$

so that negations can not be transferred across $\Vdash^v$ (as they can with $\Vdash^p$). Next observe that

$$(\mathcal{A}, \alpha) \Vdash^v P \rightarrow \square^3 P$$

for the only elements x with $x \Vdash P$ are $x = a$ and $x = c$, and each path of length three from one of these elements returns to its starting element. However, it is easy to construct a different valuation β or $\mathcal{A}$ for which

$$\mathrm{not}[(\mathcal{A}, \beta) \Vdash^v \neg P \to \square^3 P]$$

(for instance, let $\beta(P) = \{d\}$.) Thus

$$\mathrm{not}[\mathcal{A} \Vdash^u \neg P \to \square^3 P].$$

Finally, as we noted before, we have

$$\mathcal{A} \Vdash^u \square^4 P \leftrightarrow \square P.$$

Sometimes simple properties of a structure $\mathcal{A}$ ensure that it models certain formulas. For example, consider the four properties reflexivity, transitivity, being pathetic, and denseness of a relation used in Chapter 3, Section 3.4. The following result should be compared to Proposition 3.4.

4.6 LEMMA. *For a given structure $\mathcal{A}$, suppose a distinguished relation $\prec$ is, respectively,*

(i)   Reflexive
(ii) Transitive
(iii) Pathetic
(iv) Dense.

Then, in each case, for each formula ϕ, the corresponding compound formula

(i) $\square\phi \to \phi$
(ii) $\square\phi \to \square^2\phi$
(iii) $\phi \to \square\phi$
(iv) $\square^2\phi \to \square\phi$

is modelled by $\mathcal{A}$.

Proof. For instance, suppose that $\prec$ is transitive and, for an arbitrary valuation on $\mathcal{A}$ and element a, suppose that

$$a \Vdash \square\phi.$$

To show that $a \Vdash \square^2\phi$, consider elements $c \prec b \prec a$. Then, since $\prec$ is transitive, we have $c \prec a$, so that $c \Vdash \phi$, which is enough to verify case (ii) . The other three cases are verified in a similar way. ■

There are two properties of the satisfaction relations which are particularly important. These will be used later to form the basis of a proof system for modal logic. The first property provides the basic axioms.

4.7 LEMMA. *For each structure* $\mathcal{A}$

$$\mathcal{A} \Vdash^u \; [i](\theta \to \psi) \to (\, [i]\theta \to [i]\phi)$$

holds for all labels i and formulas θ and ψ.

Proof. Consider any valuation on $\mathcal{A}$ and element a of $\mathcal{A}$ such that

$$a \Vdash \; [i](\theta \to \psi) \quad , \quad a \Vdash \; [i]\theta.$$

Then, for each element $x \prec_i a$, we have

$$x \Vdash \theta \to \psi \quad , \quad x \Vdash \theta$$

so that $x \Vdash \psi$, and hence $a \Vdash \; [i]\psi$, which gives the required result. ∎

The final result of this section will eventually provide the rules of inference for modal proof systems.

4.8 LEMMA. *For each valued structure $(\mathcal{A}, \alpha)$ the implication*

$$(\mathcal{A}, \alpha) \Vdash^v \phi \;\Rightarrow\; (\mathcal{A}, \alpha) \Vdash^v \; [i]\phi$$

holds for all labels i and formulas ϕ.

Proof. Suppose that $(\mathcal{A}, \alpha) \Vdash^v \phi$ and consider any pair of elements $x \prec_i a$. Then $x \Vdash \phi$ so that $a \Vdash \; [i]\phi$, and hence $(\mathcal{A}, \alpha) \Vdash^v \; [i]\phi$, as required. ∎

You may be tempted to think that Lemma 4.8 could be improved to

$$(\mathcal{A}, \alpha) \Vdash \phi \to [i]\phi$$

or perhaps to

$$a \Vdash \phi \quad \Rightarrow \quad a \Vdash \; [i]\phi.$$

Neither of these hold in general, and you should look for appropriate counterexamples.

4.5   Semantics for modal algebras

As explained in Chapter 3, each structure $\mathcal{A}$ is equivalent to a modal algebra based on $\mathcal{P}A$. Any semantics supported by $\mathcal{A}$ can be transferred to this modal algebra, as we now describe.

By definition, a valuation on $\mathcal{A}$ is a mapping

$$\alpha : Var \longrightarrow \mathcal{P}A.$$

The forcing relation $\Vdash$ induced by α extends α to a mapping

$$Form \longrightarrow \mathcal{P}A$$

where, for each formula ϕ, the assigned subset of A is written

$$[\![\phi]\!]_\alpha.$$

This set is a measure of how true ϕ is in $(\mathcal{A}, \alpha)$.

In the definition of $[\![\phi]\!]_\alpha$ we use the boolean operations

$$\cap, \cup, \neg$$

of union, intersection, and complementation on $\mathcal{P}A$, together with the modal operations induced by the relations $\prec_i$. It is worth comparing this definition with Definitions 1.2, 2.3, and 4.2.

4.9 DEFINITION. Let $(\mathcal{A}, \alpha)$ be a given valued structure. For each formula ϕ the subset

$$[\![\phi]\!]_\alpha$$

of A is defined by recursion on ϕ using the following clauses.

(**Const**) For the constants

$$[\![\top]\!]_\alpha = A \quad, \quad [\![\bot]\!]_\alpha = \emptyset.$$

(**Var**) For each variable P

$$[\![P]\!]_\alpha = \alpha(P).$$

($\neg$) For each formula ϕ

$$[\![\neg\phi]\!]_\alpha = \neg[\![\phi]\!]_\alpha.$$

($\wedge, \vee, \rightarrow$) For all formulas θ, ψ

$$
\begin{aligned}
[\![\theta \wedge \psi]\!]_\alpha &= [\![\theta]\!]_\alpha \cap [\![\psi]\!]_\alpha \\
[\![\theta \vee \psi]\!]_\alpha &= [\![\theta]\!]_\alpha \cup [\![\psi]\!]_\alpha \\
[\![\theta \rightarrow \psi]\!]_\alpha &= \neg[\![\theta]\!]_\alpha \cup [\![\psi]\!]_\alpha
\end{aligned}
$$

(i) For each label i and formula ϕ

$$[\![\,[i]\phi\,]\!]_\alpha = [i]([\![\phi]\!]_\alpha)$$

where $[i]$ is the modal operation on $\mathcal{P}A$ induced by $\prec_i$. ∎

It is clear from this definition that for the most part the distinguishing subscript on

$$[\phi]_\alpha$$

is playing no useful purpose. We will therefore omit it in future unless this could lead to confusion (e.g. when there are two valuations around).

The two notions $\Vdash$ and $[\cdot]$ are connected in an obvious way. The proof of this is entirely routine and will be left as an exercise.

4.10 PROPOSITION. *For each valued structure* $(\mathcal{A}, \alpha)$ *the equivalence*

$$a \Vdash \phi \quad \Leftrightarrow \quad a \in [\phi]$$

holds for all elements a of $\mathcal{A}$ and formula ϕ.

Note how this shows that

$$(\mathcal{A}, \alpha) \Vdash^v \phi \quad \Leftrightarrow \quad [\phi] = A$$

which can be useful in certain situations.

This rephrasing of the semantics provides a convenient way of evaluating substitution instances.

Recall that a substitution is a map

$$\sigma : Var \longrightarrow Form$$

which extends to a map

$$Form \longrightarrow Form$$
$$\phi \longmapsto \phi^\sigma.$$

These should be compared with the maps α and $\phi \mapsto [\phi]_\alpha$; the only difference between the two cases is that $Form$ is the target set for substitutions whereas $\mathcal{P}A$ is the target set for valuations.

The question to consider is how can we determine the value

$$[\phi^\sigma]_\alpha$$

without first performing the substitution and then applying $[\cdot]_\alpha$. The method is reminiscent of the way we dealt with iterated substitutions.

4.11 DEFINITION. For each substitution σ and valuation α, let $\alpha * \sigma$ be the valuation given by

$$(\alpha * \sigma)(P) \quad = \quad [P^\sigma]_\alpha$$

for each variable P. ∎

The following result is the analogue of Exercise 2.5(a) of Chapter 2.

4.12 THEOREM. *For each substitution σ, valuation α (on a structure), and formula ϕ we have*

$$[\![\phi^\sigma]\!]_\alpha \quad = \quad [\![\phi]\!]_\beta$$

where $\beta = \alpha * \sigma$.

Proof. This follows by recursion on ϕ. The various steps are entirely routine. For instance, for the step across a box we know that

$$(\,[i]\phi)^\sigma \quad = \quad [i]\phi^\sigma$$

so that

$$[\![(\,[i]\phi)^\sigma]\!]_\alpha = [\![\,[i]\phi^\sigma]\!]_\alpha = [i]([\![\phi^\sigma]\!]_\alpha) = [i][\![\phi]\!]_\beta = [\![\,[i]\phi]\!]_\beta$$

where the third equality follows by the induction hypothesis and the others by Definition 4.9 (for both α and β). ∎

This result has a consequence for the satisfaction relation ⊩ which, at times, can be very useful.

4.13 PROPOSITION. *Suppose ψ is a substitution instance of the formula ϕ. Then the implication*

$$\mathcal{A} \Vdash^u \phi \quad \Rightarrow \quad \mathcal{A} \Vdash^u \psi$$

holds for all structures $\mathcal{A}$.

Proof. Let σ be a substitution such that $\psi = \phi^\sigma$, and suppose $\mathcal{A} \Vdash \phi$. Consider any valuation α on $\mathcal{A}$. We require $[\![\psi]\!]_\alpha = A$. Let $\beta = \alpha * \sigma$. Then

$$[\![\psi]\!]_\alpha = [\![\phi^\sigma]\!]_\alpha = [\![\phi]\!]_\beta = A$$

where the last equality holds since $\mathcal{A} \Vdash^u \phi$. ∎

As a simple application of this result we may combine it with Proposition 4.3 to obtain the following.

4.14 COROLLARY. *Suppose ψ is an instance of a propositional tautology. Then ψ holds in all structures.*

4.6 Exercises

4.1 Consider the 10 essentially different transition structures on two elements listed in Exercise 3.1 of Chapter 3. You are invited to determine which of the standard shapes D,T,B,4,... are modelled by these various structures. The table below gives a partial list of this information. You should check these result and fill in the other details.

	D	T	B	4	5	P	Q	R	G	L	M
(1)	×	×	.	✓	✓	.	✓	✓	✓	.	×
(2)	.	.	✓	✓	.	✓	.	✓	✓	×	.
(4)	.	✓	✓	.	✓	✓	✓	.	.	×	✓
(5)	×	×	.	✓	×	×	.	.	×	.	×
(7)	×	.	×	✓	.	.	×	.	×	×	×
(8)	.	×	×	.	✓	×	.	✓	×	×	.
(11)	✓	✓	.	✓	✓	×	×	✓	.	×	✓
(13)	✓	.	✓	.	.	×	✓	.	×	.	.
(14)	.	×	✓	×	×	.	×	.	.	×	×
(16)	✓	✓	✓	✓	.	×	×	.	×	×	×

4.2 Consider the three element monomodal transition structures of Exercise 3.2. Show that there are at least two non-isomorphic such structures which model none of the shapes D,T,...,M.

4.3 We may use the various order relations on $\mathbf{N}$ to define four different transition structures

$$\mathcal{N} = (\mathbf{N}, \longrightarrow)$$

where

$$(a) \quad x \longrightarrow y \;\Leftrightarrow\; x < y$$
$$(b) \quad x \longrightarrow y \;\Leftrightarrow\; x \leq y$$
$$(c) \quad x \longrightarrow y \;\Leftrightarrow\; x > y$$
$$(d) \quad x \longrightarrow y \;\Leftrightarrow\; x \geq y$$

for $x, y \in \mathbf{N}$. The following table indicates whether or not some of the standard formulas hold in $\mathcal{N}$. You are invited to check this information and fill in the rest of the table.

	D	T	B	4	5	P	Q	R	G	L	M
(a)	✓	.	×	✓	×	×	×	.	✓	×	.
(b)	.	✓	×	✓	×	.	×	✓	✓	.	×
(c)	×	×	×	.	×	×	×	.	×	.	.
(d)	✓	✓	.	✓	.	×	.	✓	.	×	✓

4.4 Each of the sets

$$A := \mathbf{N}, \mathbf{Z}, \mathbf{Q}, \mathbf{R}$$

gives us a monomodal structure $\mathcal{A} = (A, \longrightarrow)$ where

$$a \longrightarrow b \;\Leftrightarrow\; a < b$$

for each $a, b \in A$. Let us write $\mathcal{N}, \mathcal{Z}, \mathcal{Q}, \mathcal{R}$ for these structures.

For each formula ϕ consider the compound formulas

$$\mathrm{T}(\phi) := \Box\phi \to \phi \qquad\qquad \mathrm{U}(\phi) := \Box(\phi \to \Box\phi) \to \phi$$
$$\mathrm{L}(\phi) := \Box\mathrm{T}(\phi) \to \Box\phi \qquad\qquad \mathrm{V}(\phi) := \Box\mathrm{U}(\phi) \to \Box\phi$$
$$\mathrm{S}(\phi) := \Diamond\Box\phi \to \mathrm{L}(\phi) \qquad\qquad \mathrm{W}(\phi) := \Diamond\Box\phi \to \mathrm{V}(\phi).$$

Show the following

(i) $\mathcal{N} \Vdash^p S(\phi)$ (ii) $\mathcal{N} \Vdash^p W(\phi)$
(iii) $\mathcal{Z} \Vdash^p S(\phi)$ (iv) $\mathcal{Z} \Vdash^p W(\phi)$
(v) $\neg[\mathcal{Q} \Vdash^p S(P)]$ (vi) $\neg[\mathcal{Q} \Vdash^p W(\phi)]$
(vii) $\neg[\mathcal{R} \Vdash^p S(\phi)]$ (viii) $\neg[\mathcal{R} \Vdash^p W(\phi)]$

where ϕ is an arbitrary formula and P is an arbitrary variable. For (v - viii) consider any $D \subseteq A$ where both D and $A - D$ are dense in A, and let P be true on $D \cup (1, \infty)$.

4.5 Show that for each transitive (monomodal) structure $\mathcal{A}$ both

$$\mathcal{A} \Vdash^u \ \Box \Diamond^2 \phi \leftrightarrow \Box \Diamond \phi \ , \ \mathcal{A} \Vdash^u \ (\Box \Diamond)^2 \phi \leftrightarrow \Box \Diamond \phi$$

hold for all formulas ϕ.

4.6 Since our modal language contains the constants $\bot$ and $\top$, it is possible to construct formulas which contain no variables. Such a formula is called a *sentence*. For instance, in the monomodal case, the sentences are

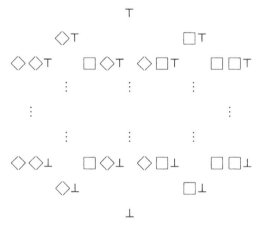

together with all the boolean and modal combinations of these. Sentences do not need a valuation in order to acquire a truth value. Thus, given a sentence ϕ, a structure $\mathcal{A}$, and an element a, if

$$a \Vdash \phi$$

holds for some valuation on $\mathcal{A}$ then it holds for all valuations on $\mathcal{A}$.

Sentences can be used to capture some information about chains in a structure.

Let us restrict ourselves to the monomodal case, and let $\mathcal{A}$ be a given structure. For each $n \in \mathsf{N}$ and elements $a, b \in A$ let

$$a \xrightarrow{n} b \tag{4.2}$$

mean there is a chain

$$a = a_0 \longrightarrow a_1 \longrightarrow \cdots \longrightarrow a_n = b$$

of length n between a and b. In particular, $a \xrightarrow{0} a$ holds vacuously. Let

$$a \xrightarrow{n} \checkmark \quad , \quad a \xrightarrow{n} \times$$

mean, respectively,

there is some b ,   there is no b

for which (4.2) holds. It is also useful to say

$$a \text{ can see } b \;\Leftrightarrow\; a \longrightarrow b$$
$$a \text{ is blind} \;\Leftrightarrow\; a \longrightarrow \times.$$

(The confusion between the use of '$\xrightarrow{n}$' as an iterated transition and a labelled transition is quite deliberate.)

(a) Show that

(i) $a \Vdash \Diamond^n \top \;\Leftrightarrow\; a \xrightarrow{n} \checkmark$

(ii) $a \Vdash \Box^n \bot \;\Leftrightarrow\; a \xrightarrow{n} \times.$

Under what conditions do

$$a \Vdash \Box^n \top \quad , \quad a \Vdash \Diamond^n \bot$$

hold?

(b) Show that

(i) $a \Vdash \Box \Diamond \bot \;\Leftrightarrow\; a$ is blind

(ii) $a \Vdash \Diamond \Box \bot \;\Leftrightarrow\; a$ can see a blind element

(iii) $a \Vdash \Box \Diamond \top \;\Leftrightarrow\; a$ can see no blind elements

(iv) $a \Vdash \Diamond \Box \top \;\Leftrightarrow\; a$ is not blind

hold.

(c) Find sentences which express the following.

(i) Every element seen by a is blind.

(ii) The element a can see a non-blind element.

(iii) Every element seen by a is either blind or can see a non-blind element.

(iv) The element a can see a non-blind element which can see only blind elements.

4.7 We say a set Φ of sentences is *independent* if for each $\phi \in \Phi$ there is a structure $\mathcal{A}$ and an element a of $\mathcal{A}$ such that

- for each $\psi \in (\Phi - \{\phi\})$, $a \Vdash \psi$
- $a \Vdash \neg\phi$.

This exercise exhibits an infinite independent set of sentences.

For each $m \in \mathsf{N}$ let $W(m)$ be the sentence

$$\Diamond^{m+2}\top \rightarrow \Box^{m+1}\Diamond\top.$$

Let $\mathcal{A}_m$ be the structure with $2m+4$ elements $a, \dots, b, \dots, c$ where

$$a \xrightarrow{m+1} b \quad , \quad a \xrightarrow{m+2} c$$

and where the chains between a and b and a and c are disjoint apart from the common source a.

(a) Draw a picture of $\mathcal{A}_0, \mathcal{A}_1, \mathcal{A}_2$, and $\mathcal{A}_3$.

(b) Show that for each $l < m < n$ both

$$a \Vdash \Box^{l+1}\Diamond\top \quad , \quad a \Vdash \Box^{n+2}\bot$$

hold in $\mathcal{A}_m$.

(c) Show that in $\mathcal{A}_m$

$$a \Vdash \neg W(m)$$

holds, whereas

$$a \Vdash W(k)$$

holds for all $k \neq m$.

(d) Show that the set $\{W(m) \mid m \in \mathsf{N}\}$ is independent.

4.8 Consider the monomodal structure $\mathcal{N} = (\mathsf{N}, \longrightarrow)$ where

$$x \longrightarrow y \Leftrightarrow x \leq y + 1$$

for all $x, y \in \mathsf{N}$. This is called the *recession structure*. To determine the truth values of formulas in $\mathcal{N}$ observe that each $X \subseteq \mathsf{N}$ falls into exactly one of the following five types.

(1) $X = \mathsf{N}$.

(2) There is some $a \in \mathbf{N}$ with $[a+1, \infty] \subseteq X \subseteq \{a\}'$ (where $(\cdot)'$ is the complement operation.

(3) There is some $a \in X$ with $X \subseteq [0, a]$.

(4) $X = \emptyset$.

(5) X is neither finite nor cofinite.

Note that types (2,3) are complementary, as are types (1,4). Type (5) is self complementary.

(a) Verify the following tables of values for these five types.

X	$\Box x$	$\Diamond X$	$\Box^2 X$	$\Diamond^2 X$	$\Box\Diamond X$	$\Diamond\Box X$
(1)	N	N	N	N	N	N
(2)	$[a+2, \infty]$	N	$[a+3, \infty]$	N	N	N
(3)	$\emptyset$	$[0, a+i]$	$\emptyset$	$[0, a+i]$	$\emptyset$	$\emptyset$
(4)	$\emptyset$	$\emptyset$	$\emptyset$	$\emptyset$	$\emptyset$	$\emptyset$
(5)	$\emptyset$	N	$\emptyset$	N	N	$\emptyset$

(b) Determine which of the standard formulas are modelled by $\mathcal{N}$.

(c) Which of the following formulas

   (i) $\Box(\Box\phi \to \Box\psi) \vee \Box(\Box\psi \to \Box\phi)$

   (ii) $\Box(\Box^2\phi \to \Box^3\phi) \to (\Box\phi \to \Box^2\phi)$

   (iii) $\Box(\Box(\phi \to \Box\phi) \to \Box^3\phi) \to \Box^4\phi$

are modelled by $\mathcal{N}$?

4.9 Consider two structures $\mathcal{A}$ and $\mathcal{B}$ of the same signature. We say $\mathcal{B}$ is a *substructure* of $\mathcal{A}$ and write

$$\mathcal{B} \subseteq \mathcal{A} \qquad\qquad (4.3)$$

if $B \subseteq A$ and the transition relations of $\mathcal{B}$ are the restrictions of the corresponding relations of $\mathcal{A}$, i.e. for each label i and elements $b, y \in B$

$$b \overset{i}{\longrightarrow} y \text{ holds in } \mathcal{B} \iff b \overset{i}{\longrightarrow} y \text{ holds in } \mathcal{A}.$$

Note that every non-empty subset of A is the carrier of a substructure of $\mathcal{A}$.

Given (4.3), each valuation α on $\mathcal{A}$ restricts to a valuation β on $\mathcal{B}$ defined by

$$b \in \beta(P) \iff b \in \alpha(P)$$

for each variable P and $b \in B$. We then say that the valued structure $(\mathcal{B}, \beta)$ is a substructure of $(\mathcal{A}, \alpha)$ and write

$$(\mathcal{B}, \beta) \subseteq (\mathcal{A}, \alpha) \tag{4.4}$$

We say $\mathcal{B}$ or $(\mathcal{B}, \beta)$ is a *generated substructure* of $\mathcal{A}$ or $(\mathcal{A}, \alpha)$ and write

$$\mathcal{B} \subseteq_g \mathcal{A} \quad \text{or} \quad (\mathcal{B}, \beta) \subseteq_g (\mathcal{A}, \alpha)$$

if the appropriate one of (4.3) or (4.4) holds and, for each label i and elements $b \in B$ and $a \in A$,

$$b \xrightarrow{i} a \;\Rightarrow\; a \in B$$

holds.

(a) Show that for a given valued structure $(\mathcal{A}, \alpha)$

   (i) each non-empty subset B of A carries a substructure $(\mathcal{B}, \beta)$ of $\mathcal{A}$,

   (ii) for each element $a \in A$ there is a smallest generated substructure $\mathcal{B} = \mathcal{A}(a)$ of $\mathcal{A}$ which contains a.

(b) Hence show that if $(\mathcal{B}, \beta) \subseteq_g (\mathcal{A}, \alpha)$ then

$$(\mathcal{B}, \beta, b) \Vdash \phi \;\Leftrightarrow\; (\mathcal{A}, \alpha, b) \Vdash \phi$$

for all $b \in B$ and formulas ϕ.

(c) For an arbitrary formula ϕ let $\{\phi\}^*$ be the set of all formulas $[i]\phi$ for a compound label i. Show that for each valued structure $(\mathcal{A}, \alpha)$ and element a of $\mathcal{A}$, the equivalence

$$(\mathcal{A}, \alpha, a) \Vdash^p \{\phi\}^* \;\Leftrightarrow\; (\mathcal{B}, \beta) \Vdash^v \phi$$

holds where $\mathcal{B} = \mathcal{A}(a)$.

4.10 Let $\mathcal{A}$ be any structure of an arbitrary signature, and let $A^\vee$ be the set of ultrafilters $p, q, r, \ldots$ on A. (Recall that an ultrafilter on A is a set p of subsets of A with certain appropriate closure properties.) For each $a \in A$ let

$$a^\vee = \{X \in \mathcal{P}A \mid a \in X\}$$

be the principal ultrafilter on a.

For each label i let $\xrightarrow{i}$ be the relation on $A^\vee$ given by

$$p \xrightarrow{i} q \;\Leftrightarrow\; (\forall X \in \mathcal{P}A)[\,[i]X \in p \Rightarrow X \in q]$$

for $p, q \in a^\vee$, where $[i]$ is the modal operation on $\mathcal{P}A$ corresponding to the label i.

This gives us a structure $\mathcal{A}^\vee$ based on $A^\vee$ which is called the *ultrafilter extension* of the structure $\mathcal{A}$.

For a given valuation α on $\mathcal{A}$ let $\alpha^\vee$ be the valuation on $\mathcal{A}^\vee$ given by

$$p \in \alpha^\vee(P) \;\Leftrightarrow\; \alpha(P) \in p$$

for each $p \in A^\vee$ and variable P. We compare the two valued structures $(\mathcal{A}, \alpha)$ and $(\mathcal{A}^\vee, \alpha^\vee)$.

(a) Show that
$$a \xrightarrow{\;i\;} b \;\Leftrightarrow\; a^\vee \xrightarrow{\;i\;} b^\vee$$
for each label i and elements $a, b \in A$.

(b) Show that
$$(\mathcal{A}^\vee, \alpha^\vee, p) \Vdash \phi \;\Leftrightarrow\; [\![\phi]\!] \in p$$
for each formula ϕ and $p \in A^\vee$.

(c) Show that

   (i) $(\mathcal{A}, \alpha, a)^\vee \Vdash^p \phi \;\Leftrightarrow\; (\mathcal{A}, \alpha, a) \Vdash^p \phi$

   (ii) $(\mathcal{A}, \alpha)^\vee \Vdash^v \phi \;\Leftrightarrow\; (\mathcal{A}, \alpha) \Vdash^v \phi$

   (iii) $\mathcal{A}^\vee \Vdash^u \phi \;\Rightarrow\; \mathcal{A} \Vdash^u \phi$ for each formula ϕ and $a \in A$.

[These results form a modal version of Łos's result from first order model theory.]

4.11 Let $\mathcal{N} = (\mathsf{N}, \longrightarrow)$ be the monomodal structure where

$$a \longrightarrow b \;\Leftrightarrow\; a < b$$

for all $a, b \in \mathsf{N}$ and consider the ultrafilter extension $\mathcal{N}^\vee$ of $\mathcal{N}$. Show the following.

(a) For each $x \subseteq \mathsf{N}$ and $p \in \mathsf{N}^\vee$ if $\Box X \in p$ then X is cofinite.

(b) For all $p, q \in \mathsf{N}^\vee$,
$$p \longrightarrow q$$
holds whenever q is non-principal.

(c) The structure $\mathcal{N}^\vee$ has some reflexive points.

(d) There are formulas ϕ such that both
$$\mathcal{N} \Vdash^u \phi \;,\quad \neg[\mathcal{N}^\vee \Vdash^u \phi]$$
hold.

4.12 In the following chapters we show that many properties of transition structures can be captured by sets of modal formulas. However, not all such properties can be captured in this way. For instance let us say a monomodal structure $\mathcal{A}$ is *good* if it is transitive, serial, and each of its elements can see a reflexive element.

(a) Show that the structure $\mathcal{N}$ of Exercise 4.11 is transitive and serial but not good, whereas its ultrafilter extension $\mathcal{N}^\vee$ is good.

(b) Show that there is no set of formulas Γ such that a structure models Γ precisely when it is good.

Chapter 5

Correspondence theory

5.1 Introduction

In Chapter 3 I claimed that modal logic should be viewed as a tool for describing and analysing properties of structures (that is, of labelled transition structures). How can this be, and how effective is this tool? The kind of simple things we might want to know about a relation are whether it is reflexive, symmetric, transitive, confluent, etc. We may want to know whether one relation is included in another, or is the converse of another, or whether one relation can be decomposed as the composite of two others, etc. We may want to know more complicated things like whether a relation is well-founded, or whether one relation is the *-closure of another.

Remarkably, these and many other properties are characterized by quite simple modal formulas. It is this characterizing ability which makes modal logic such a powerful tool. Once it is understood, it can be seen that modal logic is a quite extensive part of full second order logic, and it is the ability to capture second order properties which gives it is power.

5.2 Some examples

As an illustration of the kind of thing we are going to do we begin with a quite simple example of a correspondence result. In this result we focus on one particular label with its associated relation $\prec$ and connective $\square$.

5.1 PROPOSITION. *For each structure $\mathcal{A}$ the conditions*

(i) *The distinguished relation $\prec$ is reflexive.*
(ii) *For each formula ϕ, $\mathcal{A} \Vdash^u \square\phi \to \phi$.*
(iii) *For some variable P, $\mathcal{A} \Vdash^u \square P \to P$.*

are equivalent.

61

Before we prove this result, we make a few remarks. There are many correspondence results, and all have the same form as Proposition 5.1. A structural property (i) is shown to be equivalent to the modelling of a certain family of formulas (ii). These formulas are all the substitution instances of a certain set of basic formulas (iii). Thus the implication (ii) ⇒ (iii) is trivial, and (iii) ⇒ (ii) is an application of Proposition 4.13. The implication (i) ⇒ (ii) is always proved by direct verification (and is no harder than the implication (i) ⇒ (iii)). The implication (iii) ⇒ (i) follows by the use of a particular valuation chosen in a suitable way. The choice of this valuation is the only non-routine part of the proof.

Proof of Proposition 5.1. (i) ⇒ (ii). This is Lemma 4.6 (i).

(ii) ⇒ (iii). This is trivial

(iii) ⇒ (i). With the variable P given by (iii), for a fixed element a, consider any valuation α such that for each $x \in A$,

$$x \Vdash P \quad \Leftrightarrow \quad x \prec a$$

holds. Then (as in Chapter 4, Section 4.3) we have

$$a \Vdash \Box p$$

and hence, invoking (iii), we obtain $a \Vdash P$. Thus $a \prec a$, as required. ∎

Our second example has a very similar proof.

5.2 PROPOSITION. *For each structure $\mathcal{A}$ the conditions*

(i)   The distinguished relation $\prec$ is transitive.

(ii) For each formula ϕ, $\mathcal{A} \Vdash^u \Box\phi \to \Box^2\phi$.

(iii) For some variable P, $\mathcal{A} \Vdash^u \Box P \to \Box^2 P$.

are equivalent.

Proof. (i) ⇒ (ii). This is Lemma 4.6 (ii).

(ii) ⇒ (iii). This is trivial.

(iii) ⇒ (i). With the variable P given by (iii), for a fixed element a, consider any valuation α such that for each $x \in A$,

$$x \Vdash P \Leftrightarrow x \prec a.$$

Then $a \Vdash \Box P$ so that, invoking (iii), we have $a \Vdash \Box^2 P$. Thus, for all elements b and c,

$$c \prec b \prec a \quad \Rightarrow \quad c \Vdash P \quad \Rightarrow \quad c \prec a$$

which gives (i). ∎

It isn't always the same valuation which is required for the proof of (iii) ⇒ (i). For instance, consider the following.

5.3 PROPOSITION. *For each structure $\mathcal{A}$ the conditions*

(i)   The distinguished relation $\prec$ is deterministic.
(ii) For each formula ϕ, $\mathcal{A} \Vdash^u \Diamond \phi \to \Box \phi$.
(iii) For some variable P, $\mathcal{A} \Vdash^u \Diamond P \to \Box P$.

are equivalent.

Proof. (i) $\Rightarrow$ (ii). This left as an exercise.
(ii) $\Rightarrow$ (iii). This is trivial.
(iii) $\Rightarrow$ (i). Consider any elements a, b, c with

and, with the variable given by (iii), consider any valuation such that for each $x \in A$,

$$x \Vdash P \Leftrightarrow x = b.$$

Then $a \Vdash \Diamond P$ so that (iii) gives $a \Vdash \Box P$ and hence $c \Vdash P$, i.e. $c = b$, as required. ∎

There is a whole family of results of this kind with virtually the same proof. For instance all of the following properties of a structure $\mathcal{A}$ and label i fall into this class.

(a) $\mathcal{A}$ is i-serial, i.e. for each $a \in A$ there is some $b \in A$ with $b \prec_i a$.

(b) $\mathcal{A}$ is i-reflexive, i.e. the relation $\prec_i$ is reflexive.

(c) $\mathcal{A}$ is i-symmetric.

(d) $\mathcal{A}$ is i-transitive.

(e) $\mathcal{A}$ is i-euclidean, i.e. for each divergent wedge

we have $b \prec_i c$ (and $c \prec_i b$).

(f) $\mathcal{A}$ is i-pathetic.

(g) $\mathcal{A}$ is i-deterministic.

(h) $\mathcal{A}$ is i-dense.

For each of these properties you should attempt to prove the correspondence result. Of course, the problem is to find the characterizing formula, but once this has been done the proof is virtually routine.

5.3   The confluence property

There is a single result which covers all the correspondence results of the last section and many more as well. We begin our discussion of this result here and will return to it later (in Chapter 6).

We need some terminology.

Fix the labels

$$i, j, k, l.$$

These may be distinct or have repetitions among them. We say a structure $\mathcal{A}$ has the (i, j, k, l)-confluence property if for each divergent wedge of elements

$$
\begin{array}{c}
b \\
{}^{i}\nearrow \\
a \\
\searrow_{k} \\
c
\end{array}
\qquad (5.1)
$$

there is a convergent wedge

$$
\begin{array}{c}
b \\
\searrow_{j} \\
d \\
\nearrow_{l} \\
c
\end{array}
\qquad (5.2)
$$

This property subsumes all the properties mentioned in Section 5.2, but before we see how this comes about let us state and prove the correspondence result for confluence.

5.4   THEOREM. *For each structure $\mathcal{A}$ the conditions*

(i) $\mathcal{A}$ has (i, j, k, l)-confluence.
(ii) For each formula ϕ, $\mathcal{A} \Vdash^{u} \langle i \rangle [j] \phi \rightarrow [k] \langle l \rangle \phi$.
(iii) For some variable P, $\mathcal{A} \Vdash^{u} \langle i \rangle [j] P \rightarrow [k] \langle l \rangle P$.

are equivalent.

Proof. (i) $\Rightarrow$ (ii). Suppose $\mathcal{A}$ has the confluence property and that

$$a \Vdash \langle i \rangle [j] \phi$$

for some element a, formula ϕ, and valuation on $\mathcal{A}$. This hypothesis gives some $b \prec_i a$ with $b \Vdash [j]\phi$. We are required to verify that $a \Vdash [k]\langle l \rangle \phi$, so consider any element $c \prec_k a$. We then have a wedge (5.1), so the confluence property provides an element d with

$$d \prec_j b \quad , \quad d \prec_l c.$$

From the first of these we get $d \Vdash \phi$, and hence the second gives $c \Vdash \langle l \rangle \phi$ which is enough to complete the proof.

(ii) $\Rightarrow$ (iii). This is trivial.

(iii) $\Rightarrow$ (i). Consider any given wedge (5.1) and, with the variable given by (iii), consider any valuation such that

$$x \Vdash P \Leftrightarrow x \prec_j b$$

(for all $x \in A$). Then $b \Vdash [j]P$ and hence $a \Vdash \langle i \rangle [j]P$ so that, invoking (iii), we have $c \Vdash \langle l \rangle P$ and hence there is some $d \prec_l c$ with $d \Vdash P$. This last condition ensures that $d \prec_j b$, and so we have the required wedge (5.2). ∎

How does this result cover all the correspondence results of the last section, and how does the confluence property generalize the properties (a)–(h) of that section? Let us look at some cases.

(a) For a given relation $\prec$ (of the structure $\mathcal{A}$), choose the labels j and l so that $\prec_j$, $\prec_l$, and $\prec$ agree. Let i and k label equality, i.e. for each $x, y \in A$

$$y \prec_i x \Leftrightarrow x = y \Leftrightarrow y \prec_k x.$$

A divergent wedge (5.1) is then a triple of elements

$$b = a = c$$

i.e. an arbitrary element a. A convergent wedge is then given by an element d such that

$$d \prec b = a \quad , \quad d \prec c = a$$

so that confluence reduces to seriality. Note also that the corresponding formula of Theorem 5.4 becomes

$$\Box \phi \rightarrow \Diamond \phi$$

as expected.

(b) Let j and k label equality. Then the confluence property asserts that $\prec_i$ is a subrelation of $\prec_l$. In particular when $\prec_i$ also labels equality, this says that $\prec_l$ is reflexive.

(c) Let i and j label equality. Then the confluence property says that $\prec_k$ is included in the converse of $\prec_l$, i.e.

$$y \prec_k x \quad \Rightarrow \quad x \prec_l y.$$

In particular, when k and l both label $\prec$, this says that $\prec$ is symmetric.

(d) See if you can work this out for yourself, but be warned, there is a slight catch here.

(e) Let i, j, and k label $\prec$ and let l label equality. Then the confluence property says that $\prec$ is euclidean. The corresponding shape of formula is

$$\Diamond \Box \phi \to \Box \phi$$

which, by taking the contrapositive, is equivalent to the shape

$$\Diamond \phi \to \Box \Diamond \phi.$$

Alternatively we could let i, k, and l label $\prec$ and let j label equality.

You should work out the remaining cases for yourself. Also, for each particular case, it is instructive to go through the proof of Theorem 5.4 for that case. After a few of these you will begin to see what is going on.

5.4   Some non-confluence properties

It is not the case that all correspondence results are covered by Theorem 5.4; some of them are quite a bit more complicated. In this section we look at a couple of results which illustrate some of these possible extra complications.

For the first one it is convenient to have some terminology. Thus, for want of a better word, let us say a relation $\prec$ of a structure $\mathcal{A}$ is *tree-like* if

$$b \prec a \text{ and } c \prec a \quad \Rightarrow \quad b \prec c \text{ or } c \prec b$$

for all $a, b, c \in A$. The characterization of this property illustrates how more than one variable may be needed.

5.5 PROPOSITION. *For each structure $\mathcal{A}$ the conditions*

(i)   *The distinguished relation $\prec$ is tree-like.*

(ii) *For all formulas ϕ, ψ, $\mathcal{A} \Vdash^u \Box(\Box\phi \to \psi) \lor \Box(\Box\psi \to \phi)$.*

(iii) *For some pair P, Q of distinct variables, $\mathcal{A} \Vdash^u \Box(\Box P \to Q) \lor \Box(\Box$*

are equivalent.

Proof. (i) $\Rightarrow$ (ii). Suppose that $\prec$ is tree-like and that

$$\text{not}[a \Vdash \ \Box(\ \Box\phi \to \psi)]$$

for some element a, formulas ϕ and ψ, and valuation on $\mathcal{A}$. Then

$$a \Vdash \ \Diamond(\ \Box\phi \wedge \neg\psi)$$

so there is some $b \prec a$ with

$$b \Vdash \ \Box\phi \ , \quad b \Vdash \ \neg\psi.$$

Consider also any element $c \prec a$ with

$$c \Vdash \ \Box\psi.$$

We require that $c \Vdash \phi$.

Since $\prec$ is tree-like we know that either

$$b \prec c \quad \text{or} \quad c \prec b.$$

If the first of these holds then $b \Vdash \psi$, which we know is not so. Thus $c \prec b$ and hence $c \Vdash \phi$, as required.

(ii) $\Rightarrow$ (iii). This is trivial

(iii) $\Rightarrow$ (i). Suppose that $b \prec a$ and $c \prec a$, and, with the variables given by (iii) consider any valuation for which

$$x \Vdash P \Leftrightarrow x \prec b \ , \quad x \Vdash Q \Leftrightarrow x \prec c$$

(for $x \in A$). This is possible since P and Q are distinct. By construction we have

$$b \Vdash \ \Box P \ , \quad c \Vdash \ \Box Q$$

and we know, by the hypothesis, that either

$$a \Vdash \ \Box(\ \Box P \to Q) \quad \text{or} \quad a \Vdash \ \Box(\ \Box Q \to P).$$

If the first of these holds then $b \Vdash \ \Box P \to Q$ so that $b \Vdash Q$ and hence $b \prec c$. If the second holds then, by a similar argument $c \prec b$. $\blacksquare$

So far all the structural properties we have characterized have been elementary (in the sense that they are first order definable). The power of modal logic comes from its ability to deal with some non-elementary properties. We now give an example of such a property.

Before we give the characterization we need a preliminary result.

5.6 LEMMA. *For a structure $\mathcal{A}$ and variable P suppose that*

$$\mathcal{A} \Vdash^u \Box(\Box P \to P) \to \Box P.$$

Then the corresponding transition relation $\longrightarrow$ is transitive.

Proof. We make use of Proposition 4.5. Thus, consider a valuation such that for each $x \in A$,

$$x \Vdash P \iff (\forall y)[y \prec_* x \Rightarrow y \prec a]$$

where a is a fixed element and $\prec_*$ is the $*$-closure of $\prec$. Then $a \Vdash \Box(\Box P \to P)$ and hence, invoking the hypothesis, we have $a \Vdash \Box P$. But then, for all $b, c \in A$, we have

$$c \prec b \prec a \Rightarrow c \prec b \Vdash P \Rightarrow c \prec a$$

which gives the required result. ∎

A distinguished relation $\longrightarrow$ of a structure $\mathcal{A}$ is said to be *well-founded* if there is no sequence $(a_r | r < w)$ of elements with

$$a_0 \longrightarrow a_1 \longrightarrow \cdots \longrightarrow a_r \longrightarrow \cdots \qquad (r < w).$$

This property, even when combined with transitivity, is not elementary, hence the interest of the following result.

5.7 THEOREM. *For each structure $\mathcal{A}$ the conditions*

(i)   *The distinguished relation $\to$ is transitive and well-founded.*
(ii) *For each formula ϕ, $\mathcal{A} \Vdash^u \Box(\Box\phi \to \phi) \to \Box\phi$.*
(iii) *For some variable P, $\mathcal{A} \Vdash^u \Box(\Box P \to P) \to \Box P$.*

are equivalent.

Proof. (i) $\Rightarrow$ (ii). Suppose $\prec$ is well-founded and consider any element a, formula ϕ, and valuation with

$$a \Vdash \Box(\Box\phi \to \phi).$$

Consider also, any element b with

$$b \prec a \quad , \quad b \Vdash \neg\phi. \tag{5.3}$$

Then, from the position of a, we have $b \Vdash \neg\Box\phi$, which gives some element

$$c \prec b \quad \text{with} \quad c \Vdash \neg\phi.$$

Since $\prec$ is transitive this gives

$$c \prec a \quad \text{and} \quad c \Vdash \neg\phi.$$

Hence, by iterating this construction, we obtain a sequence of elements $(b_r | r < w)$ with

$$b = b_0 \longrightarrow b_1 \longrightarrow \cdots \longrightarrow b_r \longrightarrow \cdots \qquad (r < w)$$

and $b_r \Vdash \neg\phi$ for all $r < w$. Since well-foundedness obstructs such a sequence, we see there can be no initial element b satisfying (5.3). Thus $a \Vdash \neg \Diamond \neg\phi$, i.e. $a \Vdash \Box\phi$, as required.

(ii) $\Rightarrow$ (iii). This is trivial.

(iii) $\Rightarrow$ (i). Suppose (iii) holds. Then, by Lemma 5.6, the relation $\prec$ is transitive, so we must show that $\prec$ is well-founded.

By way of contradiction suppose there is a sequence $(a_r | r < w)$ with

$$a_0 \longrightarrow a_1 \longrightarrow \cdots \longrightarrow a_r \longrightarrow \cdots \qquad (r < w).$$

With the variable P given by (iii) consider any valuation such that, for $x \in A$,

$$x \Vdash \neg P \iff (\exists r < w)[x = a_r].$$

In particular, $a_1 \Vdash \neg P$ so that $a_0 \Vdash \Diamond \neg P$ and hence, invoking (iii), we have

$$a_0 \Vdash \Diamond(\Box P \wedge \neg P).$$

This gives some $x \prec a$ with

$$x \Vdash \Box P \quad \text{and} \quad x \Vdash \neg P.$$

The second of these ensures that $x = a_r$ for some $r < w$. But then, by the first we have

$$a_{r+1} \Vdash P$$

which is the required contradiction. ∎

5.5 Exercises

5.1 Consider a language with not necessarily distinct labels i, j, k, l. Below is a list of pairs of a formula shape (s) depending on an arbitrary formula ϕ, and a structural property (p). For each of these pairs, show that a structure models shape (s) precisely when it has property (p).

(a) (s) $\langle i \rangle \phi \rightarrow \phi$

  (p) i-pathetic.

(b)   (s) $\langle i \rangle \phi \rightarrow \langle j \rangle \phi$

       (p) The relation $\xrightarrow{i}$ is included in the relation $\xrightarrow{j}$.

(c)   (s) $\langle i \rangle \phi \rightarrow [j] \phi$

       (p) All wedges of the form

       have $c = b$.

(d)   (s) $\langle i \rangle \phi \rightarrow \langle j \rangle [k] \phi$

       (p) For each pair $a \xrightarrow{i} b$ there is some c with $a \xrightarrow{j} c$ such that the only x with $c \xrightarrow{k} x$ is $x = b$.

(e)   (s) $\langle i \rangle \phi \rightarrow [j] \langle k \rangle \phi$

       (p) All wedges of the form

       have $c \xrightarrow{k} b$.

(f)   (s) $\langle i \rangle \phi \rightarrow \langle j \rangle [k] \langle l \rangle \phi$

       (p) For each pair $a \xrightarrow{i} b$ there is some c with $a \xrightarrow{j} c$ such that for all d, if $c \xrightarrow{k} d$ then $d \xrightarrow{l} b$.

(g)   (s) $\langle i \rangle \phi \rightarrow [j] \langle k \rangle [l] \phi$

       (p) For each wedge

       there is some element d with $c \xrightarrow{k} d$ such that the only x with $d \xrightarrow{l} x$ is $x = b$.

5.2   Let i, j, k, l, m, n be fixed labels. For each of the following pairs, show that a structure models the shape (s) precisely when it has property (p).

(a) (s) $\langle i \rangle [j] \phi \rightarrow [k] \langle l \rangle [m] \phi$.

(p) For each wedge of elements

there is some element d with $c \xrightarrow{l} d$ such that

$$d \xrightarrow{m} x \Rightarrow b \xrightarrow{j} x$$

holds for all elements x.

(b) (s) $\langle i \rangle [j] \phi \rightarrow \langle k \rangle [l] \langle m \rangle \phi$.

(p) For each transition $a \xrightarrow{i} b$, there is some transition $a \xrightarrow{k} c$ such that for each transition $c \xrightarrow{l} d$, there is a wedge

(c) (s) $\langle i \rangle [j] \phi \rightarrow [k] \langle l \rangle [m] \langle n \rangle \phi$.

(p) For each wedge

there is a transition $c \xrightarrow{l} d$ such that for each transition $d \xrightarrow{m} e$

$$
\begin{array}{ccc}
& & b \\
& & \big\downarrow j \\
e & \longrightarrow & f \\
& n &
\end{array}
\quad .
$$

5.3 Consider a modal language with labels i, j, k, l, m, n (where these need be neither distinct nor atomic). Let K(i, j, k, l, m, n) be the generalized K-shape

$$[i]([j]\phi \rightarrow [k]\psi) \rightarrow [l]([m]\phi \rightarrow [n]\psi)$$

for arbitrary ϕ and ψ. Show that a structure $\mathcal{A}$ models K(i, j, k, l, m, n) if and only if for all elements a, b, c with

$$a \xrightarrow{l} b \xrightarrow{n} c$$

there is some element d such that

- $a \xrightarrow{i} d \xrightarrow{k} c$

- for all elements x, $d \xrightarrow{j} x \Rightarrow b \xrightarrow{m} x$

hold.

5.4 Not all correspondence results need to be proved by chosing a suitable valuation. For instance let k and l be fixed natural numbers with $k > l$. Show that for each transitive (monomodal) structure $\mathcal{A}$, the three conditions:

(i) For all formulas ϕ, $\mathcal{A} \Vdash^p \Box^k \Diamond \Box \phi \rightarrow \Box^l \Diamond \Box \phi$.

(ii) $\mathcal{A} \Vdash^u \Diamond^k \Box \bot \vee \Box^l \Diamond \top$.

(iii) For each $a \in A$ one of

  - There is some blind b with $a \xrightarrow{k} b$.
  - There is no blind b with $a \xrightarrow{l} b$.

  holds.

are equivalent.

5.5 Let k and l be fixed natural numbers. Show that for each transitive (monomodal) structure $\mathcal{A}$, the four conditions:

(i) For all formulas ϕ, $\mathcal{A} \Vdash^u \Box^k \Diamond \phi \rightarrow \Diamond^l \Box \Diamond \phi$.

(ii) For all formulas ϕ, $\mathcal{A} \Vdash^u \Box^k \Diamond \Box \phi \rightarrow \Diamond^l \Box \Diamond \phi$.

(iii) $\mathcal{A} \Vdash^u \Diamond^k \Box \bot \vee \Diamond^l \Box \Diamond \top$.

(iv) For each $a \in A$, there is some $b \in A$ such that one of

  - $a \xrightarrow{k} b$ and b is blind
  - $a \xrightarrow{l} b$ and no element seen by b is blind

  holds.

are equivalent.

5.6 The results of Exercise 5.1 can be generalized. To do this let us say a *modal operator* M is a sequence of $\langle i \rangle$ and $[i]$ for varying labels i. For instance

$$\emptyset, \ \langle i \rangle, \ [i], \ \langle i \rangle [i], \ [j] \langle i \rangle, \ \langle i \rangle [i] \langle k \rangle, \ldots$$

are all modal operators. These operators may be defined recursively by:

- $\emptyset$ is a modal operator;

- if M is a modal operator then so are

$$\langle i \rangle M \quad , \quad [i]M$$

for each label i.

Note that for each modal operator M and formula ϕ, the compound $M\phi$ is also a formula. Structural properties corresponding to the shapes

$$\langle i \rangle \phi \rightarrow M\phi$$

can be developed, but this requires some preliminary notation.

Fix a structure $\mathcal{A}$. For each modal operator M a relation

$$-\{M\}\!\!\rightarrow$$

is recursively defined on $\mathcal{A}$ using the following clauses.

For each pair $a, b \in A$

$$a -\{\emptyset\}\!\!\rightarrow b \quad \Leftrightarrow \quad a = b$$

$$a -\{\langle i \rangle M\}\!\!\rightarrow b \quad \Leftrightarrow \quad \text{There is some } x \text{ with} \quad a \overset{i}{\longrightarrow} x -\{M\}\!\!\rightarrow b$$

$$a -\{[i]M\}\!\!\rightarrow b \quad \Leftrightarrow \quad \text{For all } x, \quad a \overset{i}{\longrightarrow} x \Rightarrow x -\{M\}\!\!\rightarrow b$$

(for arbitrary i and M).

(a) Give explicit descriptions of the relations

$$-\{\langle i \rangle\}\!\!\rightarrow \qquad -\{[i]\}\!\!\rightarrow$$

$$-\{\langle i \rangle [j]\}\!\!\rightarrow \qquad -\{[i]\langle j \rangle\}\!\!\rightarrow$$

$$-\{\langle i \rangle [j] \langle k \rangle\}\!\!\rightarrow \quad -\{[i]\langle j \rangle [k]\}\!\!\rightarrow$$

on $\mathcal{A}$.

(b) For a fixed element b and variable P, let β be any valuation such that $\beta(P) = \{b\}$. Show that

$$a -\{M\}\!\!\rightarrow b \quad \Leftrightarrow \quad (\mathcal{A}, \beta, a) \Vdash^p MP$$

holds for all $a \in A$ and modal operators M.

(c) Show that for all modal operators M, and pairs a, b with $a -\{M\}\!\!\rightarrow b$, the implication

$$b \Vdash \phi \Rightarrow a \Vdash M\phi$$

holds for all valuations and formulas ϕ.

(d) Show that $\mathcal{A}$ models the shape

$$\langle i \rangle \phi \to M\phi$$

if and only if the relation $\xrightarrow{i}$ is included in the relation $\neg\{M\}\!\rightarrowtail$.

(e) Show how the results of Exercise (5.1) are particular cases of (d).

5.7   The results of this chapter and the previous exercises show that many formula shapes are equivalent to structural properties which are describable in elementary terms. This is not the case for all formula shapes. The simplest example of a non-elementary shape is given by the *McKinsey* formula

$$M(\phi) \; := \; \square \Diamond \phi \to \Diamond \square \phi$$

on an arbitrary formula ϕ. The class of models of this shape is quite weird, in particular the following result forms a basis of a proof that the class is not elementary.

Let S be a fixed countably infinite set and, as usual, let $2 = \{0, 1\}$. Let $[S \longrightarrow 2]$ be the set of functions

$$f : S \longrightarrow 2.$$

For each such function f let $\neg f$ be the complementary function given by

$$(\neg f)(x) = 1 - f(x)$$

(for $x \in S$). Let F be any subset of $[S \longrightarrow 2]$. We use this to construct a transition structure $\mathcal{A}(F)$.

Thus set

$$A(F) \; = \; \{a\} \cup S \cup (S \times 2) \cup F$$

where a is some new element. Let $\longrightarrow$ be the transition relation on $A(F)$ such that

$$a \longrightarrow x \longrightarrow (x, i) \longrightarrow (x, i) \qquad a \longrightarrow f \longrightarrow (x, f(x))$$

for all $x \in S, i \in 2, f \in F$, with no other transitions holding. Let

$$\mathcal{A}(F) \; = \; (A(F), \longrightarrow).$$

(a) Show that for each $u \in A(F)$:

(i) $u = a$ $\Leftrightarrow$ there is no $v \in A(F)$ with $v \longrightarrow u$

(ii) $u \in S \times 2$ $\Leftrightarrow$ $u \longrightarrow u$

(iii) $u \in S$ $\Leftrightarrow$ there are precisely two $v \in A(F)$ with $u \longrightarrow v$

(iv) $u \in F$ $\Leftrightarrow$ there are at least three $v \in A(F)$ with $u \longrightarrow v$

(b) Show that

(i) $(x, i) \Vdash \phi \leftrightarrow \Box \phi$ (ii) $(x, i) \Vdash \phi \leftrightarrow \Diamond \phi$
(iii) $x \Vdash M(\phi)$ (iv) $x \Vdash \Box \Diamond \phi \rightarrow \Box^2 \phi$
(v) $f \Vdash M(\phi)$ (vi) $f \Vdash \Box \Diamond \phi \rightarrow \Box^2 \phi$

for all appropriate x, i, f and ϕ, and valuations on $\mathcal{A}(F)$.

(c) For arbitrary $g : S \longrightarrow 2$ with $\neg g \notin F$ consider the valuation γ on $\mathcal{A}(F)$ where, for some variable P,

$$y \Vdash P \Leftrightarrow (\exists x \in S)[y = (x, g(x))]$$

for each $y \in A(F)$.

   (i) Show that $a \Vdash \Box \Diamond P$.
   (ii) Hence show that if $\mathcal{A}(F)$ models $M(P)$ then $g \in F$.

(d) Suppose the set is closed under $\neg(\cdot)$. Show that $\mathcal{A}(F)$ models the shape M if and only if $F = [S \longrightarrow 2]$.

5.8 Consider the following choice principle.

($*$) Suppose that $\longrightarrow$ is a transitive relation on the set X such that

$$(\forall x \in X)(\exists y \in X)[x \longrightarrow y \wedge x \neq y].$$

Then there are sets Y, Z with

$$Y \cap Z = \emptyset \quad , \quad Y \cup Z = X$$

and such that

$$(\forall x \in X)(\exists y \in Y, z \in Z)[x \longrightarrow y \wedge x \longrightarrow z]$$

holds.

This is a version of the Axiom of Choice. In some restricted situations non-elementary properties can become elementary.

(a) Using ($*$), show that a transitive structure $\mathcal{A} = (A, \longrightarrow)$ models McK-insey's axiom if and only if: For each $a \in A$ there is some $b \in A$ with $a \longrightarrow b$ and such that
$$b \longrightarrow x \Rightarrow x = b$$
holds for all $x \in A$.

(b) Can you prove ($*$)?

Chapter 6

The general confluence result

6.1 Introduction

In Chapter 5 we obtained several particular correspondence results covering such structural properties as

(a) Seriality
(b) Reflexivity
(c) Symmetry
(d) Transitivity
(e) $\cdots$
$\vdots$.

I also indicated how these results are all particular instances of a more general result using a confluence property. However, you may have noticed that the confluence result given in Chapter 5 (namely Theorem 5.4) is not quite general enough. To make that result cover such properties as transitivity and denseness we need to fudge the distinction between a single label and a sequence of labels. In this chapter we rework Theorem 5.4 in a generality great enough to do the job properly.

As usual we fix a signature I. We must concern ourselves with sequences of labels, for example

$$i \quad := \quad i(1), i(2), \cdots, i(p).$$

This is a sequence of length p (where $p \in \mathbb{N}$) consisting of label $i(1)$ followed by label $i(2)$ followed by ... and finishing with the label $i(p)$. These atomic labels need not be distinct. We also allow the case $p = 0$, in which case i is the empty sequence and is written $\emptyset$.

Given such a sequence i and elements a and b of a structure $\mathcal{A}$ we let

$$a \xrightarrow{\;i\;} b$$

mean there are elements $x_0, x_1, x_2, \ldots, x_p$ with

$$a = x_0 \xrightarrow{i(1)} x_1 \xrightarrow{i(2)} x_3 \xrightarrow{i(3)} \cdots \xrightarrow{i(p)} x_p = b.$$

In particular, when $p = 0$,

$$a \xrightarrow{\emptyset} b \quad \text{means} \quad a = b.$$

Similarly, for a formula ϕ we let

$$[\mathsf{i}]\phi$$

abbreviate

$$[1]\,[2]\cdots[p]\phi$$

where the indexes $1, 2, \cdots, p$ indicate the appropriate label to use. We also use a similar convention for

$$\langle \mathsf{i} \rangle.$$

In particular, both

$$[\emptyset]\phi \quad \text{and} \quad \langle \emptyset \rangle \phi$$

are just ϕ.

To ensure that you understand these conventions you should show that

$$a \Vdash [\mathsf{i}]\phi$$

holds precisely when:

For each element b with $a \xrightarrow{\mathsf{i}} b$ we have $b \Vdash \phi$.

These notations and conventions allow us to manipulate with i as though it were a single label.

To deal with the general confluence result we consider four such sequences of labels. Thus for the remainder of this chapter we fix four natural numbers

$$p \quad , \quad q \quad , \quad r \quad , \quad s$$

together with four sequences of labels as follows.

$$
\begin{aligned}
\mathsf{i} \ &:= \ i(1), i(2), \ldots, i(p) \\
\mathsf{j} \ &:= \ j(1), j(2), \ldots, j(q) \\
\mathsf{k} \ &:= \ k(1), k(2), \ldots, k(r) \\
\mathsf{l} \ &:= \ l(1), l(2), \ldots, l(s)
\end{aligned}
$$

We use these as parameters in a structural property

$$CONF(\mathsf{i}; \mathsf{j}; \mathsf{k}; \mathsf{l})$$

and a set of formulas

$$Conf(\mathsf{i}; \mathsf{j}; \mathsf{k}; \mathsf{l}).$$

We then show these are the two matching components of a correspondence result.

6.2 The structural property

We say a structure $\mathcal{A}$ has the property

$$CONF\,(\mathsf{i};\mathsf{j};\mathsf{k};\mathsf{l})$$

if for each divergent wedge of elements

$$(6.1)$$

there is a convergent wedge

$$(6.2)$$

In particular, for the cases where

$$p = q = r = s = 1$$

this property $CONF\,(\mathsf{i};\mathsf{j};\mathsf{k};\mathsf{l})$ reduces to the (i,j,k,l)-confluence property of Chapter 5. Other particular cases cover the properties (a),(b),(c), ... of that chapter (and many more).

Let us look at all of the cases where each of p, q, r, and s is either 0 or 1. There are 16 such cases which we may represent as follows.

(0)	$\emptyset;\emptyset;\emptyset;\emptyset$	(1)	$\emptyset;\emptyset;\emptyset;l$	(2)	$\emptyset;\emptyset;k;\emptyset$
(3)	$\emptyset;j;\emptyset;\emptyset$	(4)	$i;\emptyset;\emptyset;\emptyset$	(5)	$\emptyset;\emptyset;k;l$
(6)	$\emptyset;j;\emptyset;l$	(7)	$i;\emptyset;\emptyset;l$	(8)	$\emptyset;j;k;\emptyset$
(9)	$i;\emptyset;k;\emptyset$	(10)	$i;j;\emptyset;\emptyset$	(11)	$\emptyset;j;k;l$
(12)	$i;\emptyset;k;l$	(13)	$i;j;\emptyset;l$	(14)	$i;j;k;\emptyset$
(15)	$i;j;k;l$				

Here (0) is the case where $p = q = r = s = 0$; (1) is the case where $p = q = r = 0$ and $s = 1$; (7) is the case where $p = 1$, $q = r = 0$, and $s = 1$; (12) is the case where $p = 1$, $q = 0$, and $r = s = 1$, etc.

Let's spell out the details of some of these cases.

$CONF\,(0)$ This property is vacuous in the sense that it is enjoyed by all structures. It has, therefore, limited fascination.

$CONF(1)$ This says that for each element a and elements b and c with $a = b = c$, there is some element d with

$$a = b = c \xrightarrow{l} d \quad , \quad a = b = d.$$

In other words it says that the relation $\xrightarrow{l}$ is reflexive.

$CONF(2)$ This says that for all elements a, b, and c with

$$a \xrightarrow{k} c \quad \text{and} \quad a = b$$

there is an element d with

$$a = b = c = d.$$

In other words the relation $\xrightarrow{k}$ is pathetic.

$CONF(5)$ For all elements a, b, and c with

$$a = b \quad \text{and} \quad a \xrightarrow{k} c$$

there is some element d with

$$b = d \quad \text{and} \quad c \xrightarrow{l} d$$

i.e. for all elements a and c

$$a \xrightarrow{k} c \quad \Rightarrow \quad c \xrightarrow{l} a.$$

In other words the relation $\xrightarrow{k}$ is included in the converse of $\xrightarrow{l}$. In the particular case where $k = l$ this says that $\xrightarrow{k}$ is symmetric.

$CONF(6)$ For all elements a, b and c with

$$a = b = c$$

there is an element d with

$$a \xrightarrow{j} d \quad \text{and} \quad a \xrightarrow{l} d$$

in particular both $\xrightarrow{j}$ and $\xrightarrow{l}$ are serial.

$CONF(8)$ For all elements a, b, and c with

$$a = b \quad \text{and} \quad a \xrightarrow{k} c$$

there is an element d with

$$a \xrightarrow{j} d \quad \text{and} \quad c = d$$

i.e. $\xrightarrow{k}$ is included in $\xrightarrow{j}$.

CONF(11)   For each pair of elements a and c with

there is a triangle

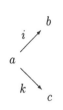

(for some d). Even with $j = k = l$ this property has not occurred before.

CONF(12)   For each wedge

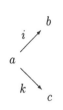

we have $c \xrightarrow{l} b$. When $i = k = l$ this says that $\xrightarrow{l}$ is euclidean.

CONF(15)   This is the confluent property of Chapter 5.

All of these properties are, in fact, instances of the original confluence property. Let us now look at some of the new properties encompassed by $CONF(i; j; k; l)$ (i.e. where at least one of $p, q, r,$ or s is two or more).

$CONF(i, i; \emptyset; \emptyset; i)$ For all elements $a, x,$ and b with

$$a \xrightarrow{i} x \xrightarrow{i} b$$

we have

$$a \xrightarrow{i} b$$

i.e. the relation $\xrightarrow{i}$ is transitive.

$CONF(i; \emptyset; \emptyset; i, i)$ For all elements a and b with

$$a \xrightarrow{i} b$$

there is some element d with

$$a \xrightarrow{i} d \xrightarrow{i} b$$

i.e. the relation $\xrightarrow{i}$ is dense.

$CONF(i,i\,;\,j\,;\,\emptyset\,;\,l,l)$ For all elements a, x, and d with

$$a \xrightarrow{\;i\;} x \xrightarrow{\;i\;} b$$

there are elements y and d with

$$b \xrightarrow{\;j\;} d \quad \text{and} \quad a \xrightarrow{\;l\;} y \xrightarrow{\;l\;} d.$$

This is not a commonly occurring property.

$CONF(i\,;\,j\,;\,k,k\,;\,l,l)$ For elements a, b, c, and x with

$$a \xrightarrow{\;i\;} d \quad \text{and} \quad a \xrightarrow{\;k\;} x \xrightarrow{\;k\;} c$$

there are elements d and y with

$$b \xrightarrow{\;j\;} d \quad \text{and} \quad c \xrightarrow{\;l\;} y \xrightarrow{\;l\;} d.$$

Again this is not a commonly occurring property.

This list of examples should convince you that these generalized confluence properties cover many (but not all) structural properties you may wish to use.

Finally, for this section, you should observe that the confluence property $CONF(\mathsf{i};\mathsf{j};\mathsf{k};\mathsf{l})$ is just the same as the property $CONF(\mathsf{k};\mathsf{l};\mathsf{i};\mathsf{j})$.

6.3   The set of formulas

Extending the notation of the previous sections let

$$Conf(\mathsf{i};\mathsf{j};\mathsf{k};\mathsf{l})$$

be the set of all formulas

$$\langle \mathsf{i} \rangle\, [\mathsf{j}]\phi \;\rightarrow\; [\mathsf{k}]\, \langle \mathsf{l} \rangle \phi$$

for arbitrary ϕ. For instance, the following shapes correspond to the examples (0 - 15) given in Section 6.2.

(0)	ϕ	$\rightarrow$	ϕ		
(1)	ϕ	$\rightarrow$	$\langle l \rangle \phi$		
(2)	ϕ	$\rightarrow$	$[k]\phi$		
(3)	$[j]\phi$	$\rightarrow$	ϕ		
(4)	$\langle i \rangle \phi$	$\rightarrow$	ϕ		
(5)	ϕ	$\rightarrow$	$[k]\langle l \rangle \phi$		
(6)	$[j]\phi$	$\rightarrow$	$\langle l \rangle \phi$		
(7)	$\langle i \rangle \phi$	$\rightarrow$	$\langle l \rangle \phi$		
(8)	$[j]\phi$	$\rightarrow$	$[k]\phi$		
(9)	$\langle i \rangle \phi$	$\rightarrow$	$[k]\phi$		
(10)	$\langle i \rangle [j]\phi$	$\rightarrow$	ϕ		
(11)	$[j]\phi$	$\rightarrow$	$[k]\langle l \rangle \phi$		
(12)	$\langle i \rangle \phi$	$\rightarrow$	$[k]\langle l \rangle \phi$		
(13)	$\langle i \rangle [j]\phi$	$\rightarrow$	$\langle l \rangle \phi$		
(14)	$\langle i \rangle [j]\phi$	$\rightarrow$	$[k]\phi$		
(15)	$\langle i \rangle [j]\phi$	$\rightarrow$	$[k]\langle l \rangle \phi$		

Many of these are refined versions of the shapes D,T,B, ... For instance, compare

(2) with P	(3) with T	(5) with B	(6) with D
(9) with Q	(12) with 5	(15) with G.	

Notice also that, by taking the contrapositive, the shape (4) is equivalent to the shape

$$\phi \rightarrow [i]\phi$$

(which is a version of(2)), shape (7) is equivalent to the shape

$$[l]\phi \rightarrow [i]\phi$$

(which is a version of the shape(8)), shape (10) is equivalent to the shape

$$\phi \rightarrow [i]\langle j \rangle \phi$$

(a version of (5)), etc.

The remaining four examples of Section 6.2 correspond to the following shapes

$$\langle i \rangle^2 \phi \;\; \rightarrow \;\; \langle i \rangle \phi$$
$$\langle i \rangle \phi \;\; \rightarrow \;\; \langle i \rangle^2 \phi$$
$$\langle i \rangle^2 [j]\phi \;\; \rightarrow \;\; \langle l \rangle^2 \phi$$
$$\langle i \rangle [j]\phi \;\; \rightarrow \;\; \langle k \rangle^2 \langle l \rangle^2 \phi$$

the first two of which are contrapositive variants of shapes 4 and R.

Finally, you should observe that, by taking contrapositives, the shape

$$Conf(i;j;k;l)$$

is equivalent to the shape

$$Conf(k;l;i;j).$$

However, no instance of $Conf(i;j;k;l)$ can produce the shape

$$\Box \Diamond \phi \rightarrow \Diamond \Box \phi.$$

This has a deep significance.

6.4   The result

In this section we prove the appropriate extension of Theorem 5.4. In fact the proof of the greater result is virtually the same as the lesser result.

6.1 THEOREM. *For each structure $\mathcal{A}$ the conditions*

(i) $\mathcal{A}$ has property $CONF(i;j;k;l)$.
(ii) $\mathcal{A}$ is a model of $Conf(i;j;k;l)$.
(iii) For some variable P, $\mathcal{A} \Vdash^u \langle i \rangle [j]P \rightarrow [k]\langle l \rangle P$.

are equivalent.

Proof. (i) $\Rightarrow$ (ii). Suppose $\mathcal{A}$ has $CONF(\mathsf{i};\mathsf{j};\mathsf{k};\mathsf{l})$ and that

$$a \Vdash \langle \mathsf{i} \rangle [\mathsf{j}] \phi$$

for some element a, formula ϕ, and valuation on $\mathcal{A}$. This hypothesis gives some element b with

$$a \xrightarrow{\mathsf{i}} b \quad \text{and} \quad b \Vdash [\mathsf{j}] \phi.$$

We are required to verify that

$$a \Vdash [\mathsf{k}] \langle \mathsf{l} \rangle \phi.$$

Consider any element c with

$$a \xrightarrow{\mathsf{k}} c.$$

Property $CONF(\mathsf{i};\mathsf{j};\mathsf{k};\mathsf{l})$ now produces an element d with

$$b \xrightarrow{\mathsf{j}} d \quad \text{and} \quad c \xrightarrow{\mathsf{l}} d.$$

From the first of these we get $d \Vdash \phi$, and hence the second gives $c \Vdash \langle \mathsf{l} \rangle \phi$, which is enough to complete the proof.

(ii) $\Rightarrow$ (iii). This is trivial.

(iii) $\Rightarrow$ (i). Consider any given wedge (6.1) and, with the variable P given by (iii), consider any valuation such that

$$x \Vdash P \Leftrightarrow b \xrightarrow{\mathsf{j}} x$$

(for all $x \in A$). Then $b \Vdash [\mathsf{j}]P$ and hence $a \Vdash \langle \mathsf{i} \rangle [\mathsf{j}]P$ so that, invoking (iii), we have $c \Vdash \langle \mathsf{l} \rangle P$. This gives us some element d with

$$c \xrightarrow{\mathsf{l}} d \quad \text{and} \quad d \Vdash P$$

which is enough to construct the required wedge (6.2). ∎

6.5   Exercises

6.1 Write down confluence formulas which capture the following structural properties.

(a) For each configuration $a \xrightarrow{i} b$ there is an element c with $b \xrightarrow{j} c$.

(b) For each configuration $a \xrightarrow{i} b$ there is an element c with $b \xrightarrow{j} c$ and $a \xrightarrow{k} c$.

(c) For each element a there is an element b with $a \xrightarrow{i} b$ and $b \xrightarrow{j} a$.

(d) For each configuration

$$a \xrightarrow{i} b \xrightarrow{j} c$$

we have $c \xrightarrow{k} a$.

(e) For each configuration

$$a \xrightarrow{i} b \xrightarrow{j} c$$

we have $a \xrightarrow{k} c$.

(f) For each configuration

$$a \xrightarrow{i} b \xrightarrow{j} c \xrightarrow{k} d$$

there is an element x with

$$a \xrightarrow{l} x \xrightarrow{m} d.$$

6.2 Below is a list of pairs of configurations L (given on the left) and R (given on the right). For each such pair consider the structural property: For each configuration L there is a configuration R. These are not confluence properties but can be captured by modal formulas. Write down appropriate formulas.

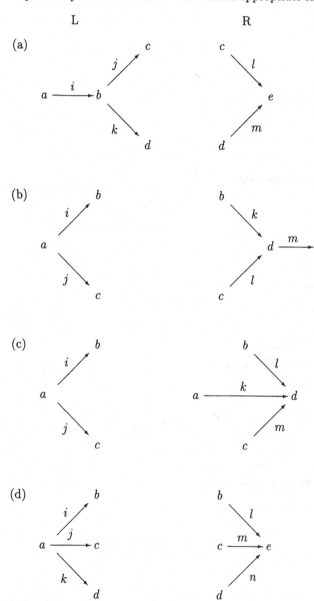

6.3 Let i, l, m, n be labels and let

$$j(1), j(2), \ldots$$
$$k(1), k(2), \ldots$$

be finite sequences of labels of the same length. For corresponding labels

$$j = j(p) \quad , \quad k = k(p)$$

taken from these sequences, and each formula ϕ_p, let

$$\langle p \rangle [p] \phi_p \quad \text{abbreviate} \quad \langle j \rangle [k] \phi_p.$$

Find a correspondence result for the shape

$$[i] (\bigwedge \{ \langle p \rangle [p] \phi_p \mid p = 1, 2, \ldots \} \to [l] \langle m \rangle (\langle n \rangle \top \wedge \bigwedge \{ \phi_p \mid p = 1, 2, \ldots \}))$$

for arbitrary formulas $\phi_1, \phi_2, \ldots$

Part III

Proof theory and completeness

This part includes all the proof theoretic machinery and results presented in this book. First, in the short motivating Chapter 7, various semantic consequence relations are introduced. Then, in Chapter 8, the notion of a *standard formal system* is developed. Such a system is given by a set of axioms and has a proof structure based on modus ponens and necessitation (a rule designed to cope with the box connectives). Once developed, this proof theoretic machinery has to be justified (in the sense that it has to be shown to be correct and powerful enough). This is done by proving a *completeness theorem*. Chapter 9 contains a completeness result which is applicable to all standard systems, and hence can be regarded as rather superficial. The proof of this result is important for the method used is applicable in many other situations. Chapter 10 contains a more refined completeness result which is widely, but not generally, applicable. This kind of completeness was first developed by Kripke and it was this advancement which brought modal logic out of the dark ages.

Chapter 7

Some consequence relations

7.1 Introduction

At this point it might be worth your while to re-read the survey of Propositional Logic, i.e. Chapter 1. There you will see that after the mechanics of 2-valuations have been set up, a semantic consequence relation

$$\Phi \models \phi$$

is defined which makes precise the informal notion

'ϕ is a logical consequence of Φ'.

The definition of this relation $\models$ makes use of various higher order notions (such as the *set of all* 2-valuations).

The programme is then continued by defining a companion relation

$$\Phi \vdash \phi$$

which is entirely combinatorial in nature. The validity of an instance of this relation can be verified by exhibiting a certain finite sequence of symbols – a witnessing deduction – which acts as a certificate for the instance. Furthermore, the legality of these certificates can be tested mechanically. The whole process is entirely finitistic.

The interaction between these two consequence relations is then explored. The soundness result, i.e. that

$$\Phi \vdash \phi \quad \Rightarrow \quad \Phi \models \phi$$

follows almost immediately from the definitions involved. However, the adequacy result, i.e. the converse

$$\Phi \models \phi \quad \Rightarrow \quad \Phi \vdash \phi$$

takes a little bit of work (which involves an essential use of a maximizing argument).

The combination of soundness and adequacy, i.e. the completeness result

$$\Phi \models \phi \quad \Leftrightarrow \quad \Phi \vdash \phi$$

should be seen as demonstrating that the fundamental notion $\models$ has a finitistic and mechanizable construction.

We now wish to attempt a similar programme for modal logic. We will see that here things are not as simple.

7.2 Semantic consequence

At first sight the appropriate definition of the notion

' the modal formula ϕ is a "logical" consequence of the set Φ of modal hypotheses'

seems straight forward. It should hold when every model of Φ is automatically a model of ϕ. However, there are at least two problems with this.

Firstly, it is not at all clear what the appropriate notion of 'model' should be. Should it be an unadorned structure, a valued structure, a pointed valued structure, or perhaps something else. There is no 'correct' answer to this problem; the several different possibilities have to be explored.

Secondly, it is not clear which modal formulas should be regarded as 'logically' valid and can thus serve as a basis for 'logical consequence'. Again there is no one answer to this; all possibilities have to be considered.

There are also several other problems which occur with modal consequence, but let us not worry about these just yet. Let us begin at the beginning.

Recall that modal structures come in three kinds k; unadorned, valued, or (valued and) pointed. Also let me remind you that at this stage we are working with an arbitrary modal language (of signature I).

7.1 DEFINITION. Let k be a kind (i.e. unadorned, valued, or valued and pointed). Then for each set of formulas Ψ and formula ϕ the relation

$$\Psi \models^k \phi$$

holds precisely when each k-structure which is a model of Ψ is also a model of ϕ. ∎

These three relations are connected in an obvious way.

7.2 PROPOSITION. *For each set of formulas Ψ and formula ϕ, both the implications*

$$\Psi \models^p \phi \quad \Rightarrow \quad \Psi \models^v \phi \quad \Rightarrow \quad \Psi \models^u \phi$$

hold.

Proof. Suppose first that

$$\Psi \models^p \phi$$

and consider any valued structure $(\mathcal{A}, \alpha)$ which models Ψ. We wish to show that $(\mathcal{A}, \alpha)$ models ϕ. To this end consider any element a of $\mathcal{A}$. Then $(\mathcal{A}, \alpha, a)$ is a valued pointed model of Ψ and hence, by the supposition, we have $a \Vdash \phi$. Since a is arbitrary this gives

$$(\mathcal{A}, \alpha) \Vdash^v \phi$$

as required.

This proves the first implication and the second follows by a similar argument. ∎

In general these two implications are not reversible. For instance, for any formula ϕ we have

$$\phi \models^v \Box \phi$$

but, for any variable P it is easy to find a valued pointed model of

$$P, \neg \Box P$$

so that

$$\phi \models^p \Box \phi$$

need not hold. You may now like to look for examples showing that

$$\models^v \quad \text{and} \quad \models^u$$

are distinct.

Note that the above example shows that $\models^v$ does not have the Deduction Property, i.e.

$$\Psi, \theta \models^v \phi \quad \Rightarrow \quad \Psi \models^v \theta \rightarrow \phi$$

need not hold (since $\models^v \theta \rightarrow \Box \theta$ does not hold in general). A similar example shows that $\models^u$ also fails to have the Deduction Property. Note, however, that $\models^p$ does have the Deduction Property.

7.3 The problem

We can now begin to see how modal logic is much more complex than non-modal logic. In particular we already have several unanswered questions:

- Which of the consequence relations $\models^k$ has a combinatorial characterization?

- How can we delineate the range of the 'logically valid' modal formulas, and how does this effect the consequence relation?

• What are the ramifications of the lack of the Deduction Property for $\models^v$ and $\models^u$?

Some of these questions will be addressed in the next two chapters where the decision made are justified by proving a couple of completeness results.

To end this chapter we prove a simple result which will form the essence of a soundness result.

7.3 LEMMA. *For all sets of formulas* Ψ *and formulas* θ, ϕ *the implications*

(*Base*) $\phi \in \Psi \quad \Rightarrow \quad \Psi \models^v \phi$

$$(MP) \quad \left. \begin{array}{c} \Psi \models^v \theta \to \phi \\[2mm] \Psi \models^v \theta \end{array} \right\} \Rightarrow \Psi \models^v \phi$$

(*N*) $\Psi \models^v \phi \quad \Rightarrow \quad \Psi \models^v [i]\phi$

hold (for each label i).

Proof. The (Base) case is trivial.
For the (MP) case observe that for each valued structure $(\mathcal{A}, \alpha)$ and formulas θ and ϕ, if

$$(\mathcal{A}, \alpha) \Vdash^v \theta \to \phi \quad \text{and} \quad (\mathcal{A}, \alpha) \Vdash^v \theta$$

then

$$(\mathcal{A}, \alpha) \Vdash^v \phi$$

(and, in fact, such an implication holds pointwise).
Finally, the (N) case follows from Lemma 4.8. ∎

7.4   Exercises

7.1 The three consequence relations $\models^k$ can be subsumed under one more general relation. Thus, for sets Θ, Ψ, Φ of formulas and a formula ϕ let

$$\Theta, \Psi, \Phi \models \phi$$

mean that for each pointed, valued structure $(\mathcal{A}, \alpha, a)$, if

$$\mathcal{A} \text{ models } \Theta \quad , \quad (\mathcal{A}, \alpha) \text{ models } \Psi \quad , \quad (\mathcal{A}, \alpha, a) \text{ models } \Phi$$

then $(\mathcal{A}, \alpha, a) \Vdash \phi$.

(a) Show that

(p) $\Phi \models^p \phi \Leftrightarrow \emptyset, \emptyset, \Phi \models \phi$

(v) $\Psi \models^v \phi \Leftrightarrow \emptyset, \Psi, \emptyset \models \phi$

(u) $\Theta \models^v \phi \Leftrightarrow \Theta, \emptyset, \emptyset \models \phi$.

(b) Show that

   (i) $\Theta, \Psi, \Phi \models \phi \Leftrightarrow \Theta, \Psi, \Psi \cup \Phi \models \phi$

   (ii) $\Theta, \Psi, \Phi \models \phi \Leftrightarrow \Theta, \Theta \cup \Psi, \Phi \models \phi$

and hence, if convenient, we may assume that

$$\Theta \subseteq \Psi \subseteq \Phi$$

whenever we use this relation.

(c) Show that if

$$\Theta \subseteq \Theta' \quad , \quad \Psi \subseteq \Psi' \quad , \quad \Phi \subseteq \Phi'$$

then

$$\Theta, \Psi, \Phi \models \phi \Rightarrow \Theta', \Psi', \Phi' \models \phi.$$

(d) Deduce Proposition 7.2.

(e) Show that

(i) $\qquad \phi \in \Phi \Rightarrow \Theta, \Psi, \Phi \models \phi$

(ii) $\left.\begin{array}{l}\Theta, \Psi, \Phi \models \theta \to \phi \\ \Theta, \Psi, \Phi \models \theta\end{array}\right\} \Rightarrow \Theta, \Psi, \Phi \models \phi$

(iii) $\qquad \Theta, \Psi, \Psi \models \phi \Rightarrow \Theta, \Psi, \Psi \models \phi$

and hence deduce Lemma 7.3.

(f) Find an example where both

$$\Theta, \Psi, \Phi \models \phi \quad , \quad \neg[\Theta, \Psi, \Phi \models [i]\phi]$$

hold.

7.2 For each set Φ of formulas let Φ^* be the closure of Φ under $[i]$ for arbitrary labels i. Thus each member of Φ^* has the form

$$[i]\theta$$

for some compound label i and $\theta \in \Phi$. Show that for each formula ϕ, the three conditions

(i) $\Phi^* \models^p \phi$, (ii) $\Phi^* \models^v \phi$, (iii) $\Phi \models^v \phi$

are equivalent. When proving the implication (iii)$\Rightarrow$(i), you may find it useful to use the properties of generated substructures as given in Exercise 4.9 of Chapter 4.

Chapter 8

Standard formal systems

8.1 Introduction

As I explained in Chapter 1, the principal aim of (non-modal) propositional logic is to give a syntactic description of the semantic consequence relation $\models$. This is done by setting up a formal system controlling a proof theoretic consequence relation $\vdash$ whose operational properties mimic (we hope) those of $\models$. The success of this programme culminates in the proof of the completeness theorem asserting that

$$\Phi \vdash \phi \qquad \Leftrightarrow \qquad \Phi \models \phi$$

for appropriate sets of formulas Φ and formulas ϕ.

We now wish to carry out a similar programme for modal logic. This is not entirely straight forward for, as we saw in Chapter 7, there are many different modal semantic consequence relations, and it is not at all clear which of these we should isolate for proof theoretic analysis.

The honest approach to this is to state at the outset which semantic consequence relations we are interested in, and then design an appropriate formal system. We won't do this. Our approach (which, in fact, is the usual approach) will be to first construct a proof theoretic consequence relation (or rather, a whole class of such consequence relations, all of a similar kind), and then see which, if any, of the semantic consequence relations have been captured.

A second difference between modal and modal-free logic has a deeper significance. In the non-modal case there are several different possible styles of proof systems (Hilbert, Natural, Sequence, ...). One of the important achievements of propositional logic (and some of its non-modal enrichments) is that these different styles are intertranslatable. (The deduction property and the cut elimination property are two of the tools used in these translations.)

In modal logic the situation is much more delicate; there are significant technical problems (some of which have not yet been solved) to be faced when translating one style into another.

We step over these problems by choosing a proof style which best suits our purpose, namely a Hilbert style of system.

For us a formal system S with its associated consequence relation $\vdash_S$ is determined by the following data.

• An acceptable set S of logical axioms.

• A set $\mathcal{R}$ of rules of inference.

The precise definition of these notions will be given later, but once they have been fixed we may construct the associated consequence relation

$$\Phi \vdash_S \phi$$

between sets of formulas Φ and formulas ϕ in a standard way. Thus this relation holds precisely when there is a witnessing formal deduction consisting of a finite sequence

$$\phi_0, \phi_1, \ldots, \phi_n$$

of formulas generated from Φ using S and finishing with the formula ϕ.

Each formal system S is determined by its two components S (the axioms) and $\mathcal{R}$ (the rules). We will fix the rules once and for all so, for us, the system S is determined solely by the choice of axioms S.

By its very nature a proof theoretic system is concerned with syntactic manipulations in a certain formal language. It is thus important that we know what is a part of that language and what are merely convenient devices for talking about the language. Therefore, for clarity, I will restate the meaning of 'the modal language of signature I'.

8.1 CONVENTION. *For an index set I, the modal language of signature I has as its basic symbols $\bot, \top, \wedge, \vee, \rightarrow, \neg$, and $[i]$ for each $i \in I$. (In particular, $\leftrightarrow$ and $\langle i \rangle$ are not part of the language.)* ■

8.2 Formal systems defined

We need to make precise the following notions.

• An acceptable set of axioms S.

• The rules of inference $\mathcal{R}$.

• A witnessing deduction $\phi_0, \phi_1, \ldots, \phi_n$.

This we now do.

8.2 DEFINITION. An acceptable set of axioms is a set S of formulas which

- contains all (modal-free) tautologies,

- contains all formulas of the shape

$$(\text{K}) \qquad [i](\theta \to \psi) \to ([i]\theta \to [i]\psi)$$

for all formulas θ, ψ and labels i,

- is closed under substitution. ■

Each set $\mathcal{A}$ of formulas generates a smallest acceptable set of axioms $\mathcal{S}$. Namely, let $\mathcal{A}^+$ be the set formed by adding to $\mathcal{A}$ all tautologies and all formulas (K) and then let $\mathcal{S}$ be the set of substitution instances of $\mathcal{A}^+$. This gives us a convenient way of describing particular sets of axioms, for we need only give a generating set $\mathcal{A}$. There are a host of examples of such sets, many of which are formed from various combinations of the shapes D, T, B, ... of Chapter 2. We look at some of these examples in the next section.

8.3 DEFINITION. The rules of inference are modus ponens

$$(\text{MP}) \qquad \frac{\theta \quad \theta \to \phi}{\phi}$$

and necessitation

$$(\text{Ni}) \qquad \frac{\phi}{[i]\phi}$$

for each label $i \in I$. ■

The meaning of these rules will become clear after we have described the notion of a formal deduction.

8.4 DEFINITION. Let $\mathcal{S}$ be the set of axioms determining a given formal system S and let Φ be an arbitrary set of formulas (the hypothesis set).

(a) A witnessing S-deduction from Φ is a finite sequence

$$\phi_0, \phi_1, \ldots, \phi_n$$

of formulas such that for each index $0 \le r \le n$ one of the following holds.

(**hyp**) The formula ϕ_r is an hypothesis, i.e. a member of Φ.

(**ax**) The formula ϕ_r is an axiom, (i.e. a member of $\mathcal{S}$).

(**mp**) The formula ϕ_r is obtained by (MP) from two earlier formulas, i.e. there are indexes $t, s < r$ with

$$\phi_t = (\phi_s \to \phi_r).$$

(n) The formula ϕ_r is obtained by (Ni) from an earlier formula, i.e. there is an index $s < r$ with

$$\phi_r = [i]\phi_s$$

(for some label $i \in I$).

(b) A formula ϕ is an S-consequence of Φ

$$\Phi \vdash_S \phi$$

if there is a witnessing S-deduction from Φ whose final term is ϕ. ∎

A better understanding of these notions will be obtained by looking at some particular examples of formal deductions. For this purpose several such examples are given in the next section. For the time being let us look at some box (and diamond) manipulations.

8.5 LEMMA. *For each formal system S and formulas ψ and ϕ,*

$$\vdash_S \psi \to \phi \quad \Rightarrow \quad \vdash_S \Box\psi \to \Box\phi$$

holds.

Proof. Let us abbreviate $\psi \to \phi$ by θ. Then any witnessing deduction for θ can be extended to one for $\Box\psi \to \Box\phi$ as follows.

$$
\begin{array}{ll}
\quad\vdots & \\
\theta & (Hyp) \\
\Box\theta & (N) \\
\Box\theta \to (\Box\psi \to \Box\phi) & (K) \\
\Box\psi \to \Box\phi & (MP)
\end{array}
$$

Down the right hand side I have indicated the justification for each of the terms in the deduction. ∎

A useful consequence of this is concerned with equivalences.

8.6 COROLLARY. *For all formulas ψ, ϕ*

$$\vdash_S \psi \leftrightarrow \phi \quad \Rightarrow \quad \vdash_S \Box\psi \leftrightarrow \Box\phi$$

holds. In particular, if ψ and ϕ are tautologically equivalent then

$$\vdash_S \Box\psi \leftrightarrow \Box\phi$$

holds.

Recall that the diamond connective $\Diamond$ has been introduced by

$$\Diamond \phi \quad \equiv \quad \neg \, \Box \, \neg \phi.$$

Thus, using the sequence of tautological equivalences

$$\neg \Diamond \neg \phi \leftrightarrow \neg \neg \, \Box \, \neg \neg \phi \leftrightarrow \Box \, \neg \neg \phi \leftrightarrow \Box \, \phi$$

we have

$$\vdash_{\mathsf{S}} \Box \phi \leftrightarrow \neg \Diamond \neg \phi.$$

Similarly, by taking contrapositives, the above Corollary gives

$$\vdash_{\mathsf{S}} \psi \leftrightarrow \phi \quad \Rightarrow \quad \vdash_{\mathsf{S}} \Diamond \psi \leftrightarrow \Diamond \phi.$$

This shows that for most purposes $\Diamond$ can be regarded as a primitive symbol.
We conclude this section with a couple of observations.

There is some considerable confusion in the literature over the correct notion of a proof theoretic consequence relation. Most authors use a weaker version which can be described in various ways but which amounts to the following notion.

8.7 DEFINITION. For a set of formulas Φ and formula ϕ let

$$\Phi \vdash_{\mathsf{S}}^{w} \phi$$

mean there is some finite part $\phi_1, \ldots, \phi_n$ of Φ with

$$\vdash_{\mathsf{S}} \phi_1 \wedge \ldots \wedge \phi_n \to \phi$$

(where here the hypothesis set is empty). ∎

It is an easy exercise (which you should do) to show that

$$\Phi \vdash_{\mathsf{S}}^{w} \phi \quad \Rightarrow \quad \Phi \vdash_{\mathsf{S}} \phi$$

holds, however, in general, the converse is false. Note, however, that when used with an empty hypothesis set, the two relations agree, i.e.

$$\vdash_{\mathsf{S}}^{w} \phi \quad \Leftrightarrow \quad \vdash_{\mathsf{S}} \phi$$

holds for all formulas ϕ.
The important difference between $\vdash_{\mathsf{S}}^{w}$ and $\vdash_{\mathsf{S}}$ is the Deduction Property.
By construction we have

$$\Phi, \theta \vdash_{\mathsf{S}}^{w} \phi \quad \Rightarrow \quad \Phi \vdash_{\mathsf{S}}^{w} (\theta \to \phi)$$

but, in general, this is not true of $\vdash_S$. To see this note that for any formula ϕ, the rule (N) gives

$$\phi \vdash_S \Box \phi.$$

but only for special systems S (the pathetic systems) do we have

$$\vdash_S (\phi \to \Box \phi).$$

It is precisely this 'defect' of $\vdash_S$ which leads many people to consider only the weaker version $\vdash_S^w$.

The second observation concerns the monotonicity of this consequence notion. We state the relevant result but leave the proof as an exercise.

8.8 PROPOSITION. *For each pair of sets of axioms S and T with $S \subseteq T$ and for each pair of hypothesis sets Φ and Ψ with $\Phi \subseteq \Psi$, the implication*

$$\Phi \vdash_S \phi \quad \Rightarrow \quad \Psi \vdash_T \phi$$

holds for all formulas ϕ.

8.3   Some monomodal systems

In this section we look at some particular, and well known, examples of formal systems formulated in the monomodal language, i.e. the language with just one label. All of these examples are generated using various combinations of the standard shapes of formulas D, T, B, 4, 5 (as given in Chapter 2).

The first and smallest formal system K is the one whose axioms are all instances of tautologies together with all instances of the shape K. (Thus the set of axioms of K is the smallest allowed by Definition 8.2.)

Larger systems can be formed by extending the set of axioms. Thus, for instance, let

$$KD \ , \ KT \ , \ KB \ , \ K4 \ , \ K5$$

be the systems whose axioms comprise the smallest acceptable set containing the formulas

$$D \ , \ T \ , \ B \ , \ 4 \ , \ 5$$

respectively. Similarly let

$$KDT \ , \ KDB \ , \ KD4 \ , \ KD5$$
$$KTB \ , \ KT4 \ , \ KT5$$
$$KB4 \ , \ KB5$$
$$K45$$

be the systems whose axioms comprise the smallest acceptable set containing the indicated shapes. Continuing further we may form such systems as

$$KDTB \ , \ KDB4 \ , \ KB45$$

etc.

Two of these systems are particularly important and have given names. These are

$$\mathsf{S4 = KT4} \quad , \quad \mathsf{S5 = KT5}.$$

On the face of it the five shapes D, T, B, 4, and 5 give us $2^5 = 32$ different systems, however, as we will see, not all of these are distinct. There are, in fact, only 15 such systems.

How do we compare two systems and what do we mean by two systems being 'the same'? We take a pragmatic, extensional view of this.

Thus given two systems S and T (based on the sets of axioms $\mathcal{S}$ and $\mathcal{T}$) we write

$$\mathsf{S \le T}$$

if

$$\vdash_\mathsf{S} \phi \quad \Rightarrow \quad \vdash_\mathsf{T} \phi$$

holds for all formulas ϕ. It can be seen that $\mathsf{S \le T}$ implies the apparently stronger property that

$$\Phi \vdash_\mathsf{S} \phi \quad \Rightarrow \quad \Phi \vdash_\mathsf{T} \phi$$

holds for all hypothesis sets Φ and formulas ϕ.

We now agree to say that S and T are the same if both the comparisons

$$\mathsf{S \le T} \quad \text{and} \quad \mathsf{T \le S}$$

hold.

Using this notation let us look at some of the comparisons which hold between $\mathsf{K, KD, KT, \dots, KDTB45}$. These comparisons will follow from various examples of witnessing deductions.

Our first example shows how the shape D is captured by T. For an arbitrary formula ϕ let

$$\alpha = \square \neg \phi \quad , \quad \beta = \square \phi \quad , \quad \gamma = \neg \alpha = \Diamond \phi$$

(so that $\mathrm{D}(\phi)$ is $\beta \to \gamma$, $\mathrm{T}(\phi)$ is $\beta \to \phi$, and $\mathrm{T}(\neg\phi)$ is $\alpha \to \neg\phi$). The following sequence is a witnessing S-deduction for any system S whose axioms include the shape T. The justification for each formula is given on the right.

$$
\begin{array}{ll}
\alpha \to \neg\phi & (T) \\
(\alpha \to \neg\phi) \to (\phi \to \neg\alpha) & (Taut) \\
\phi \to \gamma & (MP) \\
\beta \to \phi & (T) \\
(\beta \to \phi) \to ((\phi \to \gamma) \to (\beta \to \gamma)) & (Taut) \\
(\phi \to \gamma) \to (\beta \to \gamma) & (MP) \\
\beta \to \gamma & (MP)
\end{array}
$$

This deduction immediately gives the following.

8.9 LEMMA. KD $\leq$ KT   ,   KD4 $\leq$ S4   ,   KD5 $\leq$ S5.

The next example shows how S5 captures the shape B. Some of the terms have been omitted and you should fill in these positions for yourself.

$$
\begin{array}{ll}
\Box \neg \phi \rightarrow \neg \phi & (T) \\
\quad \vdots & (Taut) \\
\phi \rightarrow \Diamond \phi & (MP) \\
\Diamond \phi \rightarrow \Box \Diamond \phi & (5) \\
\quad \vdots & (Taut) \\
\quad \vdots & (MP) \\
\phi \rightarrow \Box \Diamond \phi & (MP)
\end{array}
$$

Combining this with the previous Lemma we get the following.

8.10 LEMMA. KDB5 $\leq$ S5.

All modal systems have the following two useful derived rules of inference.

$$
\frac{\theta \rightarrow \phi}{\Box \theta \rightarrow \Box \phi} \quad , \quad \frac{\theta \rightarrow \phi}{\Diamond \theta \rightarrow \Diamond \phi}
$$

We refer to these jointly as (EN), i.e. as 'extended necessitation'. The first of these rules is justified by

$$
\begin{array}{ll}
\theta \rightarrow \phi & (Hyp) \\
\Box(\theta \rightarrow \phi) & (N) \\
\Box(\theta \rightarrow \phi) \rightarrow (\Box \theta \rightarrow \Box \phi) & (K) \\
\Box \theta \rightarrow \Box \phi & (MP)
\end{array}
$$

and the second by

$$
\begin{array}{c}
\theta \rightarrow \phi \\
\vdots \\
\neg \phi \rightarrow \neg \theta \\
\vdots \\
\Box \neg \phi \rightarrow \Box \neg \theta \\
\vdots \\
\Diamond \theta \rightarrow \Diamond \phi
\end{array}
$$

These derived rules make some deductions easier to display. For instance, working in any extension of **KB5** we have

$$\Diamond \neg \phi \rightarrow \Box \Diamond \neg \phi \qquad (5)$$

$$\vdots$$

$$\Diamond \Box \phi \rightarrow \Box \phi$$

$$\vdots \qquad\qquad\qquad (EN)$$

$$\Box \Diamond \Box \phi \rightarrow \Box \Box \phi$$

$$\Box \phi \rightarrow \Box \Diamond \Box \phi \qquad (B)$$

$$\vdots$$

$$\Box \phi \rightarrow \Box \Box \phi$$

which immediately gives the following.

8.11 LEMMA. K4 ≤ KB5 , KDB4 ≤ KDB5.

The previous example showed how, in the presence of B, the shape 4 can be captured by the shape 5. We can now do the reverse.

$$\Box \neg \phi \rightarrow \Box^2 \neg \phi \qquad (4)$$

$$\vdots$$

$$\Diamond^2 \phi \rightarrow \Diamond \phi$$

$$\vdots \qquad\qquad\qquad (EN)$$

$$\Box \Diamond^2 \phi \rightarrow \Box \Diamond \phi$$

$$\Diamond \phi \rightarrow \Box \Diamond^2 \phi \qquad (B)$$

$$\vdots$$

$$\Diamond \phi \rightarrow \Box \Diamond \phi$$

As a consequence of this we have the following.

8.12 LEMMA. K5 ≤ KB4 , KT5 ≤ KTB4.

Finally we show how to capture the shape T.

$$\Box \phi \rightarrow \Box^2 \phi \qquad (4)$$

$$\Box^2 \phi \rightarrow \Diamond \Box \phi \qquad (D)$$

$$\vdots$$

$$\Box \phi \rightarrow \Diamond \Box \phi$$

$$\neg \phi \rightarrow \Box \Diamond \phi \qquad (B)$$

$$\vdots$$

$$\Diamond \Box \phi \rightarrow \phi$$

$$\vdots$$

$$\Box \phi \rightarrow \phi$$

Combining this with several of the previous results gives us two different axiomatizations of **S5**.

8.13 THEOREM. S5 = KDB4 = KDB5.

Proof. We have

$$
\begin{aligned}
\text{S5} &= \text{KT5} \\
&\le \text{KTB4} && \text{(by Lemma 8.12)} \\
&\le \text{KDB4} && \text{(from above)} \\
&\le \text{KDB5} && \text{(by Lemma 8.11)} \\
&\le \text{S5} && \text{(by Lemma 8.10)}
\end{aligned}
$$

as required. ∎

8.4   Some polymodal systems

Historically the monomodal systems of the previous section were developed for various philosophical reasons, for instance, as attempts to formalize some of the properties of the modalities 'is necessary', 'is obligatory', 'is known', etc. Alongside these developments there were also various analyses of the properties of tenses (in natural languages) and time. This brought forth tense logic which has now become *temporal logic* and which also encompasses a much wider field of applicability.

Temporal logic is a the study of certain bimodal system (and various enrichments of these) designed to capture the flow of time. The associated transition structures have the form

$$
\mathcal{A} \;=\; (A, \xrightarrow{+}, \xrightarrow{-})
$$

where the two carried transition relations are the forward passage and the backwards passage through time. The two relations are not unconnected, however the minimal restrictions we need to put on them are as follows.

8.14 DEFINITION. A *temporal structure* is a bimodal structure

$$
\mathcal{A} \;=\; (A, \xrightarrow{+}, \xrightarrow{-})
$$

where each of the transition relations is transitive, and each is the converse of the other. ∎

Using some of the correspondence results that we have obtained we see that such structures can be isolated by a standard formal system. Thus consider the bimodal language with box operators $[-]$ and $[+]$, and in this language consider all the formulas of the following shapes.

$$
\begin{aligned}
[+]\phi &\to [+]^2\phi \\
[-]\phi &\to [-]^2\phi \\
\phi &\to [+]\langle-\rangle\phi \\
\phi &\to [-]\langle+\rangle\phi
\end{aligned}
$$

Let **TEMP** be the standard formal system axiomatized by the formulas of these above four shapes.
The following result is straight forward.

8.15 THEOREM. *A bimodal structure $\mathcal{A}$ (as above) is a temporal structure precisely when it models the system* **TEMP**.

Proof. The first two axioms ensure that the two relations are transitive, the third ensures that $\xrightarrow{+}$ is included in the converse of $\xrightarrow{-}$, and the fourth ensures that $\xrightarrow{-}$ is included in the converse of $\xrightarrow{+}$. ∎

As remarked already, temporal logic can be used to analyse some of the tense properties of natural languages. The recent development of situation theory is an attempt to analyse the more general information content of natural languages. This has thrown up another bimodal system.

Thus consider the bimodal language with box operators $[\approx]$ and $\Box$ and let **SL** (situation logic) be the formal system whose axioms are all the formulas

$$[\approx]\phi \to \phi \qquad\qquad \Box\phi \to \phi$$
$$[\approx]\phi \to [\approx][\approx]\phi \qquad\qquad \Box\phi \to \Box\Box\phi$$
$$\phi \to [\approx]\langle\approx\rangle\phi$$

and the formulas

$$[\approx]\Box\phi \to \Box[\approx]\phi$$

for arbitrary formulas ϕ. Structures for this language have the form

$$\mathcal{A} = (A, \xrightarrow{\approx}, \longrightarrow)$$

and the models of **SL** are easily characterized.

8.16 THEOREM. *A structure $\mathcal{A}$ (as above) models* **SL** *precisely when the three conditions*

- *the relation $\xrightarrow{\approx}$ is an equivalence*

- *the relation $\longrightarrow$ is a pre-ordering*

- *for each configuration $a \longrightarrow b \xrightarrow{\approx} c$ there is an element d with $a \xrightarrow{\approx} d \longrightarrow c$*

hold.

Proof. A routine application of various correspondence results. ∎

Dynamic logic is a naturally occurring polymodal logic. Furthermore in this logic the set of labels has its own algebraic structure and this leads to some quite intricate properties. At brief discussion of this logic is given in Chapter 14.

8.5   Soundness properties

In Chapter 7 we introduced three semantic consequence relations $\models^k$ (for $k = u, v, p$). These made no reference to any underlying basis of 'logically valid' modal formulas. We can now correct this omission.

8.17   DEFINITION. Let S be a standard formal system with set of axioms $\mathcal{S}$. Let k be a kind. For each set of formulas Φ and formula ϕ, the relation

$$\Phi \models^k_S \phi$$

holds precisely when each k-structure which is a model of S and of Φ is also a model of ϕ. ∎

Note that

$$\Phi \models^k_S \phi \quad \Leftrightarrow \quad \Phi \cup \mathcal{S} \models^k \phi$$

so that an analysis of $\models^k_S$ could be reduced to one of $\models^k$. However, the parameterized version $\models^k_S$ leads to a much richer theory.

We have now attached to each formal system S five consequence relations; the two proof theoretic relations

$$\vdash^w_S \quad , \quad \vdash_S$$

and the three semantic relations

$$\models^p_S \quad , \quad \models^v_S \quad , \quad \models^u_S .$$

How do these relations interact? We have already observed that

$$\Phi \vdash^w_S \phi \quad \Rightarrow \quad \Phi \vdash_S \phi$$

and a simple application of Proposition 7.2 (with $\Psi = \Phi \cup \mathcal{S}$) gives

$$\Phi \models^p_S \phi \quad \Rightarrow \quad \Phi \models^v_S \phi \quad \Rightarrow \quad \Phi \models^u_S \phi.$$

We can also add to this a soundness result.

8.18   THEOREM. *For each formal system* S, *hypothesis set* Φ, *and formula* ϕ, *the implication*

$$\Phi \vdash_S \phi \quad \Rightarrow \quad \Phi \models^v_S \phi$$

holds.

Proof. This is proved by induction on the deduction witnessing $\Phi \vdash_S \phi$. The base cases and the induction steps follow by an application of Lemma 7.3 ■

This result can be used to show that different sets of axioms give different formal systems. For instance, by Lemma 8.9 we know that $KD \leq KT$. Thus to show that the two systems are not the same it suffices to find a formula ϕ with

$$\vdash_{KT} \phi \quad , \quad \text{not}[\vdash_{KD} \phi]$$

and, by the soundness result, the second of these can be justified by showing

$$\text{not}[\models^v_{KD} \phi].$$

To this end, for an arbitrary variable P let

$$\phi = T(P) = (\,\square P \rightarrow P).$$

Trivially $\vdash_{KT} \phi$.

Now consider any structure $\mathcal{A}$ with just two points a and b and accessibility relation

$$a \longrightarrow b \longrightarrow a$$

i.e. for all points x

$$x \prec a \quad \Leftrightarrow \quad x = b \quad , \quad x \prec b \quad \Leftrightarrow \quad x = a.$$

Note that $\mathcal{A}$ is serial and hence is a model of KD.

Consider any valuation α on $\mathcal{A}$ with

$$\alpha(P) = \{b\}.$$

Then $(\mathcal{A}, \alpha)$ models KD. Also

$$a \Vdash \neg P \quad , \quad b \Vdash P$$

so that

$$a \Vdash \neg P \quad , \quad a \Vdash \square P$$

and hence

$$a \Vdash \neg \phi.$$

Thus $(\mathcal{A}, \alpha)$ does not model ϕ which gives the required result.

In the next chapter we prove a couple of completeness results; one applicable to all standard formal systems, the other applicable only to a restricted class. For both results we do not deal with the full power of the appropriate consequence relation, but only with the 'logical' case, i.e. we restrict to the case where the hypothesis set Φ is empty.

8.6 Exercises

8.1 Verify that

(i) $\vdash_K \Box(\theta \wedge \psi) \leftrightarrow \Box\theta \wedge \Box\psi$

(ii) $\vdash_K \Diamond(\theta \rightarrow \psi) \rightarrow (\Box\theta \rightarrow \Diamond\psi)$

(iii) $\vdash_K \Diamond T \rightarrow D(\phi)$

(iv) $\vdash_K \neg\Diamond\theta \rightarrow \Box(\theta \rightarrow \phi)$

(v) $\vdash_K \Box\phi \rightarrow \Box(\theta \rightarrow \phi)$

(vi) $\vdash_K (\Diamond\theta \rightarrow \Box\phi) \rightarrow \Box(\theta \rightarrow \phi)$

(vii) $\vdash_K (\Diamond\theta \rightarrow \Box\phi) \rightarrow (\Box\theta \rightarrow \Box\phi)$

(viii) $\vdash_K (\Diamond\theta \rightarrow \Box\phi) \rightarrow (\Diamond\theta \rightarrow \Diamond\phi)$

(ix) $\vdash_K \Diamond\theta \rightarrow D(\phi)$

for arbitrary formulas θ, ψ and ϕ.

8.2 Show that

(i) $\vdash_{K4} \Box\Diamond^2\phi \leftrightarrow \Box\Diamond\phi$

(ii) $\vdash_{K4} (\Box\Diamond)^2\phi \leftrightarrow \Box\Diamond\phi$

and hence verify the results of Exercise 4.5.

8.3 For the four shapes of formulas D,T,B,4 consider the $16 = 2^4$ possible extensions of the system K obtained by adding some of these shapes as axioms.

(a) Show that this gives no more than 11 different systems with inclusions as shown.

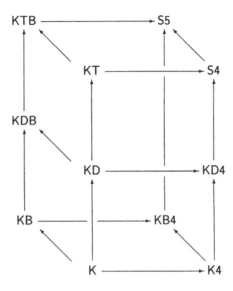

(b) By considering structures with no more than three elements, show that these 11 systems are distinct.

8.4 For each of the systems S of Exercise 8.3, consider the system S′ formed by adding 5 as a further axiom. Show that this produces no more than four new systems

$$K5, K45, KD5, KD45$$

and fit these into the diagram of Exercise 8.3.

8.5 For a given formula ϕ in a monomodal language, the *modal variants* of ϕ are the formulas $M\phi$ where M is a modal operator of the form

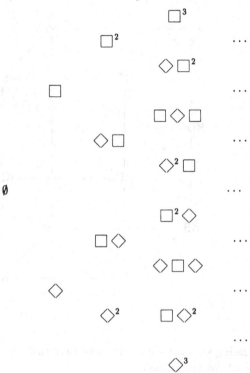

etc. In general these are all distinct and there no implications between them. However, in S4 = KT4 each formula has no more than 7 modal variants. These with some implications between them are shown in the following diagram.

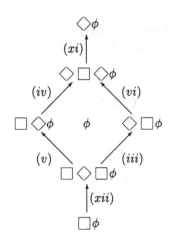

You are invited to verify these implications and to insert the two missing ones. To do this let ⊢ be ⊢$_{S4}$ and prove the following.

(a) Use T and (EN), and finally 4 to show that

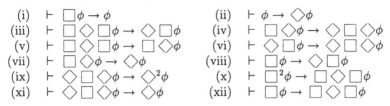

(i)   ⊢ $\Box\phi \to \phi$ (ii)   ⊢ $\phi \to \Diamond\phi$
(iii) ⊢ $\Box\Diamond\Box\phi \to \Diamond\Box\phi$ (iv)   ⊢ $\Box\Diamond\phi \to \Diamond\Box\Diamond\phi$
(v)   ⊢ $\Box\Diamond\Box\phi \to \Box\Diamond\phi$ (vi)   ⊢ $\Diamond\Box\phi \to \Diamond\Box\Diamond\phi$
(vii) ⊢ $\Box\Diamond\phi \to \Diamond\phi$ (viii) ⊢ $\Box\phi \to \Diamond\Box\phi$
(ix) ⊢ $\Diamond\Box\Diamond\phi \to \Diamond^2\phi$ (x)   ⊢ $\Box^2\phi \to \Box\Diamond\Box\phi$
(xi) ⊢ $\Diamond\Box\Diamond\phi \to \Diamond\phi$ (xii) ⊢ $\Box\phi \to \Box\Diamond\Box\phi$

hold for all formulas ϕ.

(b) Setting $\phi := \Diamond\phi$ in (a) and using T and 4, show that

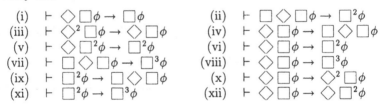

$$\vdash (\Box\Diamond)^2\phi \leftrightarrow \Box\Diamond\phi \quad , \quad \vdash (\Diamond\Box)^2\phi \leftrightarrow \Diamond\Box\phi$$

and hence each sequence of four or more modal operators collapses to three or fewer.

(c) By considering suitable models of S4, show that there is at least one formula ϕ for which the above 11 variants are distinct.

8.6 Consider the formal system K5 and let ⊢ be ⊢$_{K5}$.

(a) Verify that

(i)   ⊢ $\Diamond\Box\phi \to \Box\phi$ (ii)   ⊢ $\Box\Diamond\Box\phi \to \Box^2\phi$
(iii) ⊢ $\Diamond^2\Box\phi \to \Diamond\Box\phi$ (iv)   ⊢ $\Diamond\Box\phi \to \Box\Diamond\Box\phi$
(v)   ⊢ $\Diamond\Box^2\phi \to \Box^2\phi$ (vi)   ⊢ $\Diamond\Box\phi \to \Box^2\phi$
(vii) ⊢ $\Box\Diamond\Box\phi \to \Box^3\phi$ (viii) ⊢ $\Diamond\Box\phi \to \Box^3\phi$
(ix) ⊢ $\Box^2\phi \to \Box\Diamond\Box\phi$ (x)   ⊢ $\Diamond\Box\phi \to \Diamond^2\Box\phi$
(xi) ⊢ $\Box^2\phi \to \Box^3\phi$ (xii) ⊢ $\Diamond\Box\phi \to \Diamond\Box^2\phi$

hold for all formulas ϕ.

(b) Show that in K5 the modal variants of a formula ϕ are arranged as

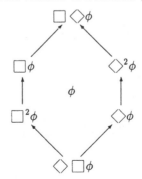

and these formulas can be distinct.

8.7   Show that in S5 each variable has precisely three modal variants and describe how these are arranged.

8.8   Let $\vdash$ be $\vdash_{\mathsf{KTB}}$. Show that for each variable P, formula ϕ and integers $m \geq n \geq 0$, both

$$\vdash \Box^m \phi \to \Box^n \phi \quad , \quad \neg[\vdash \Box^n P \to \Box^{n+1} P]$$

hold, and hence P has infinitely many modal variants in KTB.

8.9   Describe the modal variants of a variable P in K45 and in KB4.

8.10   Find temporal structures which do *not* model the following shapes.

(i) $[+]\bot$ (ii) $\langle+\rangle\top$ (iii) $[+]\phi \to \phi$ (iv) $\phi \to [$

(v) $[-]^2\phi \to [+]\phi$ (vi) $[+]^2\phi \to [+]\phi$ (vii) $\langle+\rangle\phi \to [+]\phi$ (viii) $[+]\phi -$

8.11   Which of the following are modelled by all temporal structures?

(i) $\langle+\rangle[+]\phi \to [+]\langle+\rangle\phi$ (ii) $\langle+\rangle[+]\phi \to [-]\langle-\rangle\phi$

(iii) $\langle+\rangle[-]\phi \to [-]\langle+\rangle\phi$ (iv) $\langle+\rangle[-]\phi \to [+]\langle-\rangle\phi$

8.12   The basic temporal system TEMP is too weak to capture many of the assumed properties of the passage of time. In this and the next two exercises we consider some suitable strengthenings of TEMP. For each formula ϕ let

$$[\approx]\phi \quad \text{abbreviate} \quad [-]\phi \wedge \phi \wedge [+]\phi$$

and let LINTIM be the extension of TEMP formed by the addition of the two axioms

$$[\approx]\phi \to [-][+]\phi \quad , \quad [\approx]\phi \to [+][-]\phi.$$

To understand the import of this, for each temporal structure $\mathcal{A}$, let $\approx$ be the relation on $\mathcal{A}$ given by

$$a \approx b \iff a \xrightarrow{-} b \text{ or } a = b \text{ or } a \xrightarrow{+} b$$

(for $a, b \in A$).

(a)   Show that for each model $\mathcal{A}$ of TEMP

$$a \Vdash \phi \iff \begin{cases} \text{For all } x, \\ a \approx x \Rightarrow x \Vdash \phi \end{cases}$$

holds for all valuations on $\mathcal{A}, a \in A$ and formulas ϕ.

(b)   Show that for each model $\mathcal{A}$ of TEMP the three conditions:

(i) For each formula ϕ, $\mathcal{A} \Vdash^u \, [\approx]\phi \to [-]\,[+]\phi$.

(ii) For each formula ϕ, $\mathcal{A} \Vdash^u \, \langle + \rangle\,[\approx]\phi \to [+]\phi$.

(iii) For each wedge

of elements we have $b \approx c$.

are equivalent.

(c) For an arbitrary model $\mathcal{A}$ of LINTIM:

(i) Show that the relation $\approx$ is the least equivalence relation which includes $\xrightarrow{+}$ (and $\xrightarrow{-}$).

(ii) Show that each $\approx$-equivalence class of $\mathcal{A}$ is linearly ordered by $\xrightarrow{+}$.

(d) Conversely, show that each disjoint union of linearly ordered sets provides a model for LINTIM.

8.13 This exercise continues Exercise 8.12. Each linearly ordered set $(A, <)$ produces a model of LINTIM (by interpreting $a \xrightarrow{+} b$ as $a < b$, etc). In particular

$$\mathcal{N} = (\mathsf{N}, <) \quad , \quad \mathcal{Z} = (\mathsf{Z}, <) \quad , \quad \mathcal{Q} = (\mathsf{Q}, <) \quad , \quad \mathcal{R} = (\mathsf{R}, <)$$

are models of LINTIM.

(a) Find a sentence ϕ modelled by each of $\mathcal{N}, \mathcal{Z}, \mathcal{Q}$, and $\mathcal{R}$ but for which $\neg[\vdash_{\mathsf{LINTIM}} \phi]$.

(b) Find sentences θ and ψ with

$$\mathcal{N} \Vdash^u \theta \quad , \quad \neg[\mathcal{Z} \Vdash^u \theta] \quad , \quad \neg[\mathcal{N} \Vdash^u \psi] \quad , \quad \mathcal{Z} \Vdash^u \psi.$$

(c) Find a formula shape which is modelled by both $\mathcal{Q}$ and $\mathcal{R}$ but by neither $\mathcal{N}$ nor $\mathcal{Z}$.

(d) Can you find a formula ϕ such that $\mathcal{R} \Vdash^u \phi$ but $\neg[\mathcal{Q} \Vdash^u \phi]$?

8.14 For some structures the interpretation of the two operators $\Box$ and $\Diamond$ agree, and in this case we often write $\bigcirc$ for both. Such structures are controlled by a function **next** and we think of this as a ticking clock.

(a) For a structure $\mathcal{A}$ with a transition relation $\longrightarrow$ corresponding to the modal operators $\Box$ and $\Diamond$, show that the following are equivalent.

(i) There is a function next $: A \longrightarrow A$ such that

$$a \longrightarrow b \Leftrightarrow \text{next}(a) = b$$

holds for all $a, b \in A$.

(ii) For all formulas ϕ, $\mathcal{A} \Vdash^u \Box \phi \leftrightarrow \Diamond \phi$.

(b) Let **A** be the class of structures $\mathcal{A} = (A, \longrightarrow, \overset{\bullet}{\longrightarrow})$ where $\longrightarrow$ is a ticking clock (given by the function next) and $\overset{\bullet}{\longrightarrow}$ is the $*$-closure of $\longrightarrow$. Show that **A** models each of the shapes

$$\bigcirc \phi \leftrightarrow \neg \bigcirc \neg \phi \qquad\qquad [\bullet]\phi \leftrightarrow [\bullet]^2 \phi$$

$$[\bullet]\phi \leftrightarrow \phi \wedge \bigcirc [\bullet]\phi \qquad [\bullet](\phi \to \bigcirc \phi) \to (\phi \to [\bullet]\phi)$$

(for all formulas ϕ).

8.15 Let S be any non-empty set and let Σ be any collection of non-empty subsets of S with $S \in \Sigma$. We call the elements s of S the *situations*, and we call the elements σ of Σ the *infons* (packets of information). We say a situation s *supports* an infon σ if $s \in \sigma$.

Let A be the set of supported infons i.e. the set of pairs

$$a = (s, \sigma) \quad \text{where} \quad s \in \sigma.$$

Define the relations $\overset{\approx}{\longrightarrow}$ and $\longrightarrow$ on A by

$$(s, \sigma) \overset{\approx}{\longrightarrow} (t, \tau) \quad \text{means} \qquad\qquad \sigma = \tau$$
$$(s, \sigma) \longrightarrow (t, \tau) \quad \text{means} \quad \sigma = \tau \text{ and } t \in \tau$$

and set

$$\mathcal{A} = (A, \overset{\approx}{\longrightarrow}, \longrightarrow).$$

(a) Show that $\overset{\approx}{\longrightarrow}$ is an equivalence relation, and $\longrightarrow$ is a partial ordering of A.

(b) Show that $\mathcal{A}$ models SL.

(c) Suppose that $S \in \Sigma$. Show that $\mathcal{A}$ models the confluence shape

$$\Diamond \Box \phi \to \Box \Diamond \phi$$

if and only if for each $\sigma, \tau \in \Sigma$ and $s \in \sigma \cap \tau$, there is some $\rho \in \Sigma$ with $s \in \rho \subseteq \sigma \cap \tau$.

8.16   As in Exercise 7.2 of Chapter 7, for each set Φ of formulas let Φ^* be the closure of Φ under $[i]$ for arbitrary labels i. Thus each member of Φ^* has the form

$$[i]\theta$$

for some compound label i and $\theta \in \Phi$. Let S be an arbitrary standard formal system.

(a) Show that

$$\Phi \vdash_S \Phi^*$$

and all three of the implications

$$\Phi \vdash_S^w \phi \;\Rightarrow\; \Phi^* \vdash_S^w \phi \;\Rightarrow\; \Phi \vdash_S \phi \;\Rightarrow\; \Phi^* \vdash_S \phi$$

hold for all formulas ϕ. Furthermore, show that the left hand implication is not reversible.

(b) Mimic the proof of the Deduction Theorem for propositional logic to show that the central and right hand implications of (a) are reversible.

(c) Extend the result of Exercise 7.2 to show that the three conditions

$$\Phi^* \models_S^p \phi \;\;,\;\; \Phi^* \models_S^v \phi \;\;,\;\; \Phi \models_S^v \phi \;\;,$$

are equivalent.

Chapter 9

The general completeness result

9.1 Introduction

Let S be a standard formal system and let M be a class of structures suitable for the language of S. The members of M may be unadorned, valued, or valued and pointed as the case may be, but all must be of the same kind. We say that S and M are *completely matched* if for each formula ϕ we have

$$\vdash_S \phi \quad \Leftrightarrow \quad M \Vdash^k \phi$$

where $\Vdash^k$ is the satisfaction relation appropriate for the kind of M. Such an equivalence is called a *completeness result*, where the implication $\Rightarrow$ is the *soundness* component and the implication $\Leftarrow$ is the *adequacy* component.

A completeness result of this kind is usually the solution to one of two different kinds of problems.

(a) Here we are given the system S and the problem is to find a class M which completely matches S. The reason for doing this is to analyse the properties of S in a more algebraic way, and the choice of M should take this into account.

(b) Here we are given M and the problem is to find a system S which completely matches M. The reason in doing this is to obtain a uniform way of describing the common properties of the structures in M. Thus it is desirable to make S as simple as possible.

Of course, for an arbitrarily given S or M there may be no matching partner, or there may be several. Thus proving a completeness result is not just a routine exercise. Nevertheless, such results exhibit some general features and there are some commonly used techniques. These will be described in this and the following chapters.

In this chapter we will prove a completeness result which is applicable to all standard systems S (and provides a solution to a problem of type (a)). This universal applicability means that the result is, in fact, rather weak; however the result does provide a basis for the vast majority of completeness results, in the sense that many of the proofs of these results are refinements of the universal proof.

9.2 Statement of the result

For the remainder of this chapter let S be a fixed, but arbitrary, standard formal system with axiom S. We will produce a class of valued structures which completely matches S. In fact we will produce two such classes which, in some sense, are at opposite extremes.

For one extreme let $\mathbf{M}$ be the class of all valued structures which model S (i.e. are models of S). Note that for each formula ϕ we have

$$\mathbf{M} \Vdash^v \phi \quad \Leftrightarrow \quad \models^v_S \phi$$

(for the right hand side is defined to mean the left hand side). Also, by the general soundness result, we have

$$\vdash_S \phi \quad \Rightarrow \quad \models^v_S \phi$$

so a completeness result will follow if we can prove the converse of this last implication.

For the other extreme we will construct a particular valued structure $(\mathfrak{S}, \sigma)$ – called the *canonical valued structure of* S – which models S (i.e. is a member of $\mathbf{M}$) and on its own completely matches S.

Putting these together we see that the eventual aim of this chapter is to prove the following completeness and characterization result.

9.1 THEOREM. *Let* S *be a standard formal system with canonical valued structure* $(\mathfrak{S}, \sigma)$. *For each formula* ϕ *the conditions*

(ii) $(\mathfrak{S}, \sigma) \Vdash^v \phi$
(iii) $\vdash_S \phi$
(iv) $\models^v_S \phi$

are equivalent.

(The numbering of the items of this result has been done to facilitate a comparison with a later result.)

Note that the implication (iii) $\Rightarrow$ (iv) is just soundness (which we have already proved in Chapter 8). The implication (iv) $\Rightarrow$ (ii) follows immediately we have verified that $(\mathfrak{S}, \sigma)$ models S. Much of the content of this theorem is in the implication (ii) $\Rightarrow$ (iii).

9.3 Maximally consistent sets

Intuitively a set of formulas Φ is inconsistent relative to S if the machinery of S can be used to derive a contradiction from Φ. Because of the failure of the Deduction Property we need to take a little care in making this idea precise. It turns out that the crucial property is the existence or not of formulas $\phi_1, \ldots, \phi_n$ in Φ such that

$$\vdash_S \phi_1 \wedge \ldots \wedge \phi_n \to \bot \tag{9.1}$$

The existence of such formulas means that Φ is inconsistent (relative to S), and the non-existence means that Φ is consistent. This can be made precise using the weak proof consequence relation $\vdash_S^w$.

9.2 DEFINITION.

(a) A set of formulas Φ is S-*consistent* if $\neg[\Phi \vdash_S^w \bot]$ i.e. if there are no members $\phi_1, \ldots, \phi_n$ of Φ for which (9.1) holds. Let **CON** be the collection of all such S-consistent sets.

(b) A set of formulas is *maximally* S-*consistent* if it is S-consistent but no proper extension of it is. Let **S** be the set of all such maximally S-consistent sets. ∎

In more detail, each member s of **S** is a member of **CON** and for each formula ϕ

$$s \cup \{\phi\} \in \textbf{CON} \quad \Rightarrow \quad \phi \in s$$

holds (for if $\phi \notin s$ then $s \cup \{\phi\}$ is a proper extension of s and so cannot be consistent). This maximality ensures that each $s \in \textbf{S}$ has several useful closure properties. Clearly, there is no formula ϕ such that both ϕ and $\neg\phi$ are in s (for, trivially,

$$\vdash_S \sigma \wedge \neg\phi \to \bot).$$

Also, if $\phi \notin s$ then $s \cup \{\phi\}$ is not consistent, so there is a conjunction σ of finitely many member of s with

$$\vdash_S \sigma \wedge \phi \to \bot. \tag{9.2}$$

Similarly, if $\neg\phi \notin s$ then

$$\vdash_S \tau \wedge \neg\phi \to \bot$$

for some conjunction τ of members of s. But then, using an appropriate tautology, we have

$$\vdash_S \sigma \wedge \tau \to \bot$$

which would mean that s is, in fact, inconsistent. Since this is not so, at least one of ϕ and $\neg\phi$ is in s, and hence s contains precisely one of ϕ and $\neg\phi$.

A similar argument shows that s is closed under implication. For suppose, for some formula ϕ, there is a conjunction ρ of members of s with

$$\vdash_S \rho \to \phi.$$

Then ϕ is in s, for otherwise there is a suitable conjunction σ such that (9.2) holds and hence

$$\vdash_S \sigma \wedge \rho \to \bot$$

which would mean that s is inconsistent.

By continuing in this way we may arrive at a proof of the following Proposition. (The remaining details are left as an exercise.)

9.3 PROPOSITION. *Let $s \in S$. Then*

$$\top \in s \quad , \quad \bot \notin s$$

and for all formulas θ, ψ, ϕ, the equivalences

$(\neg)$	$\neg\phi \in s$	$\Leftrightarrow$	$\phi \notin s$		
$(\wedge)$	$\theta \wedge \psi \in s$	$\Leftrightarrow$	$\theta \in s$	and	$\psi \in s$
$(\vee)$	$\theta \vee \psi \in s$	$\Leftrightarrow$	$\theta \in s$	or	$\psi \in s$
$(\to)$	$\theta \to \psi \in s$	$\Leftrightarrow$	$\theta \notin s$	or	$\psi \in s$

hold.

Note that this Proposition does not contain a clause corresponding to the connective $[i]$. The appropriate property for this will be dealt with in the next section.

The most important property of S is that it is non-empty and, in fact, has enough members to distinguish between all formulas which ought to be distinguishable. The precise result is as follows.

9.4 LEMMA. (Basic Existence Result) *For each S-consistent set of formulas Φ there is some $s \in S$ with $\Phi \subseteq s$.*

Proof. Let $(\phi_r \mid r < \omega)$ be any enumeration of all formulas, and define the sequence $(\Delta_r \mid r < \omega)$ of sets of formulas by

$$
\begin{aligned}
\Delta_0 &= \Phi \\
\Delta_{r+1} &= \begin{cases} \Delta_r \cup \{\phi_r\} & \text{if this is S-consistent} \\ \Delta_r & \text{otherwise.} \end{cases}
\end{aligned}
$$

By construction we have

$$\Phi = \Delta_0 \subseteq \Delta_1 \subseteq \cdots \subseteq \Delta_r \subseteq \cdots \qquad (r < \omega)$$

and each Δ_r is S-consistent. Let

$$s = \bigcup \{\Delta_r \mid r < \omega\}.$$

This set s is consistent. For if it isn't, then some finite part of s is inconsistent, and this finite part is included in some Δ_r, which is consistent. Thus it remains to show that s is maximal.

Consider any formula ϕ such that $s \cup \{\phi\} \in \boldsymbol{CON}$. There is at least one index $r < \omega$ such that $\phi = \phi_r$. But then

$$\Delta_r \cup \{\phi_r\} \subseteq s \cup \{\phi\}$$

so that the smaller set is also consistent, and hence

$$\phi = \phi_r \in \Delta_{r+1} \subseteq s$$

to demonstrate the required maximality of s. ■

As remarked already, this is the most important result of the whole chapter. In fact Theorem 9.1 (as is also the case with several other results) is little more than a rewording of Lemma 9.4. Almost all of what follows is a slightly elaborate exercise in symbol shuffling.

A simple consequence of the Basic Existence Result (i.e. of Lemma 9.4) is a characterization of S-derivability from a set of hypotheses.

9.5 LEMMA. *For each set of formulas* Φ *and formula* ϕ, *the equivalence*

$$\Phi \vdash_{\mathsf{S}}^{w} \phi \quad \Leftrightarrow \quad (\forall s \in \boldsymbol{S})[\Phi \subseteq s \Rightarrow \phi \in s]$$

holds.

Proof. ($\Rightarrow$). This holds since each $s \in \boldsymbol{S}$ is closed under implication.

($\Leftarrow$). The hypothesis (ii) together with the Basic Existence Result ensure that

$$\Phi \cup \{\neg\phi\} \notin \boldsymbol{CON}.$$

But then there is a conjunction τ of members of Φ such that

$$\vdash_{\mathsf{S}} \tau \wedge \neg\phi \to \bot.$$

An application of a tautology now gives (i). ■

One particular case of this Lemma is worth noting separately, namely the case $\Phi = \emptyset$.

9.6 COROLLARY. *For each formula* ϕ, *the equivalence*

$$\vdash_{\mathsf{S}} \phi \quad \Leftrightarrow \quad (\forall s \in \boldsymbol{S})[\phi \in s]$$

holds.

9.4   The canonical structure

Each structure $\mathcal{A}$ (labelled transition structure) comprises a non-empty carrying set A furnished with an appropriately indexed family of binary relations (transition relations) $\xrightarrow{i}$ (one for each label i). The canonical structure $\mathfrak{S}$ for S is such a structure based on the set S of maximally consistent sets of formulas. To complete the construction of S it remains to define the corresponding family of relations. Thus, for each index i let

$$\xrightarrow{i}$$

be the relation on S where for each $s, t \in S$,

$$s \xrightarrow{i} t$$

holds precisely when for all formulas ϕ,

$$[i]\phi \in s \quad \Rightarrow \quad \phi \in t.$$

We also use

$$t \prec_i s$$

to indicate that $s \xrightarrow{i} t$ holds. (As usual, this will help to condense certain descriptions.)

Note how the $(\neg)$ clause of Proposition 9.3 shows that $s \xrightarrow{i} t$ holds exactly when for each formula ϕ

$$\phi \in t \quad \Rightarrow \quad \langle i \rangle \phi \in s.$$

The Basic Existence Result now allows us to add to the equivalences of this Proposition.

9.7 LEMMA. *For each $s \in S$ and formula ϕ, the equivalence*

$$[i]\phi \in s \quad \Leftrightarrow \quad (\forall t \prec_i s)[\phi \in t]$$

holds.

Proof. ($\Rightarrow$) This follows immediately from the definition of $\prec_i$.
($\Leftarrow$) Consider set of formulas

$$\Psi = \{\psi \mid [i]\psi \in s\}.$$

Then the definition of $\prec_i$ and the hypothesis (the right hand side) gives

$$\Psi \subseteq t \in S \quad \Rightarrow \quad t \prec_i s \quad \Rightarrow \quad \phi \in t$$

so that Lemma 9.5 provides $\psi_1, \ldots, \psi_n \in \Psi$ with

$$\vdash_S \psi_1 \wedge \ldots \wedge \psi_n \rightarrow \phi.$$

Using the basic properties of S, this gives

$$\vdash_S \; [i]\psi_1 \wedge \ldots \wedge [i]\psi_i \rightarrow [i]\phi.$$

Since each $[i]\psi_r \in s$ and s is closed under implications, this gives $[i]\psi \in s$, as required. ∎

9.5 The canonical valuation

As we will see later, in general the unadorned canonical structure $\mathfrak{S}$ is not a model of S. However, it does model S when enriched by suitably chosen valuations. The *canonical valuation* σ on $\mathfrak{S}$ is given by

$$\sigma(P) = \{s \in \boldsymbol{S} \mid P \in s\}$$

for variables P. Equivalently, σ is such that

$$s \Vdash P \quad \Leftrightarrow \quad P \in s$$

for $s \in \boldsymbol{S}$ and variables P. This equivalence extends naturally.

9.8 LEMMA. *For each $s \in \boldsymbol{S}$ and formula ϕ, the equivalence*

$$s \Vdash \phi \quad \Leftrightarrow \quad \phi \in s$$

holds.

Proof. For each formula ϕ consider the condition

$$(\phi) \qquad (\forall s \in \boldsymbol{S})[s \Vdash \phi \Leftrightarrow \phi \in s].$$

We verify (ϕ) by induction on the complexity of ϕ.

The base case holds by the definition of σ (and since $\top \in s$ and $\bot \notin s$). The passage across the propositional connectives follow from the equivalences of Proposition 9.3. It thus remains to pass across $[i]$.

For a given $s \in \boldsymbol{S}$ and formula ϕ, using first the definition of $\Vdash$ and then the Induction Hypothesis (ϕ) followed by Lemma 9.7, we have

$$s \Vdash [i]\phi \quad \Leftrightarrow \quad (\forall t \prec_i s)[t \Vdash \phi]$$
$$\Leftrightarrow \quad (\forall t \prec_i s)[\phi \in t] \quad \Leftrightarrow \quad [i]\phi \in s$$

which gives $([i]\phi)$, and so completes the proof. ∎

As an immediate consequence of this with Corollary 9.6 we have the following.

9.9 COROLLARY. *The canonical valued structure $(\mathfrak{S}, \sigma)$ models S.*

9.6   Proof of the result

The proof of Theorem 9.1 is now very short. For instance, for each formula ϕ, Lemma 9.8 and Corollary 9.6 give

$$(\mathfrak{S}, \sigma) \Vdash^v \phi \iff (\forall s \in \boldsymbol{S})[s \Vdash \phi]$$
$$\iff (\forall s \in \boldsymbol{S})[\phi \in s] \iff \vdash_S \phi$$

which verifies the implication (ii) $\Rightarrow$ (iii). The implication (iii) $\Rightarrow$ (iv) is just soundness, and the implication (iv) $\Rightarrow$ (ii) is a consequence of Corollary 9.9.
∎

9.7   Concluding remarks

On the face of it, Theorem 9.1 gives us two solutions to problem (a) (of Section 9.1) for an arbitrary standard system S. However, these solutions are not of much practical value. On the one hand we are told we may analyse S by looking at the class of all its valued models. Unfortunately, to investigate this large class we must already have a fairly extensive knowledge of S (in which case we do not need the completeness result). On the other hand we are told we may analyse S by looking at a particular valued model. Unfortunately, this model is constructed from S and the process of determining its properties involves an analysis of S (which makes the exercise somewhat pointless).

In spite of these drawbacks, Theorem 9.1 does have some value. It does at least tell us that every standard system has a single characteristic valued model (a fact which is not at all obvious). However, to bring out its full potential, we must now refine the Theorem. We may do this by either strengthening the conclusion, or by modifying the proof to extract more information.

There are two possible lines of development.

One possibility is to try to eliminate the references to valuations, and look for a characteristic class M consisting of unadorned structures. This can be done for certain pleasantly disposed systems S (for which $\mathfrak{S}$ itself models S). It turns out that such systems S have a characteristic class M defined entirely without reference to S, and hence the completeness result does open up a genuine second line of attack on S. This case is discussed in the next chapter.

Another possibility is to look for a characteristic class which consists entirely of finite structures. This then opens up the possibility of a mechanical test for derivability within the system. (For, clearly, checking validity in a finite structure is potentially mechanizable.) Again we find there are many systems for which this approach is feasible, and these are discussed in a later chapter.

There are, of course, systems with no known completeness result (beyond that of Theorem 9.1) and there are examples of theories with various demonstrable complexities (or eccentricities). A first course in modal logic is not the place for a detailed account of these, but such examples should at least be mentioned. (Otherwise you might get an over rosy view of modal life.) A few of these more complex systems will be described later.

9.8 Exercises

9.1 Fill in the details of the proof of Proposition 9.3.

9.2 This exercise continues and makes use of the notation of and is related to the content of Exercise 7.2 and 8.16. Thus, for an arbitrary formal system S and set Φ of formulas, let $S(\Phi)$ be the set of all $s \in S$ with $\Phi^* \subseteq s$. This set $S(\Phi)$ is converted into a transition structure $\mathfrak{S}(\Phi)$ using the restriction to $S(\Phi)$ of the transition relations of $\mathfrak{S}$. Thus $\mathfrak{S}$ is the particular case $\mathfrak{S}(\emptyset)$. In the same way let σ be the restriction to $\mathfrak{S}(\Phi)$ of the canonical valuation on $\mathfrak{S}$.

(a) Show that for each $s \in S(\Phi)$, formula ϕ, and label i, the conditions

- $[i]\phi \in s$
- for each $t \in S(\Phi)$, if $s \overset{i}{\longrightarrow} t$ then $\phi \in t$

are equivalent.

(b) Show that
$$s \Vdash \phi \Leftrightarrow \phi \in s$$
holds for all $s \in S(\Phi)$ and formulas ϕ.

(c) Show that $(\mathfrak{S}(\Phi), \sigma)$ models $S \cup \Phi^*$.

(d) Show that for each formula ϕ, the conditions

(i) $\Phi^* \vdash_S^w \phi$

(ii l) $\Phi \vdash_S \phi$ (ii r) $\Phi \models_S^v \phi$

(iii l) $\Phi^* \vdash_S \phi$ (iii r) $\Phi^* \models_S^v \phi$

(iv) $(\mathfrak{S}(\Phi), \sigma)$ models ϕ

are equivalent.

Chapter 10

Kripke-completeness

10.1 Introduction

Each formal system S has associated with it three primary consequence relations

$$\vdash_S \quad , \quad \models_S^v \quad , \quad \models_S^u$$

(as well as several secondary ones). From the soundness properties we know that for each formula ϕ both the implications

$$\vdash_S \phi \quad \Rightarrow \quad \models_S^v \phi \quad \Rightarrow \quad \models_S^u \phi$$

hold. The completeness result of the previous chapter shows that the first of these implications is an equivalence and, furthermore, there is a fixed valued structure $(\mathfrak{S}, \sigma)$ such that both components are equivalent to

$$(\mathfrak{S}, \sigma) \Vdash^v \phi.$$

In this chapter we bring in the third component $\models_S^u \phi$. In general this is not equivalent to the other two, so we are interested in the following notion.

10.1 DEFINITION. A formal system S is *Kripke-complete* if

$$\vdash_S \phi \quad \Leftrightarrow \quad \models_S^u \phi$$

holds for all formulas ϕ. ■

The aim of this chapter is to show that many (if not all) of the formal systems we are interested in are Kripke-complete. In fact most of these enjoy a stronger property.

10.2 DEFINITION. A formal system S is *canonical* if its canonical (unadorned) structure $\mathfrak{S}$ is a model of S. ■

129

For instance, the smallest theory K is canonical (since every modal structure is a model of K). Later in this chapter we will generate many more examples of canonicity. Before we do that let us see why canonicity is useful.

10.3 THEOREM. *Let* S *be a canonical formal system with canonical valued structure* $(\mathfrak{S}, \sigma)$. *For each formula* ϕ *the conditions*

> *(i)* $\mathfrak{S} \Vdash^u \phi$
>
> *(ii)* $(\mathfrak{S}, \sigma) \Vdash^v \phi$
>
> *(iii)* $\vdash_{\mathsf{S}} \phi$
>
> *(iv)* $\models^v_{\mathsf{S}} \phi$
>
> *(v)* $\models^u_{\mathsf{S}} \phi$

are equivalent. In particular, each canonical system is Kripke-complete.

Proof. The implication (i) $\Rightarrow$ (ii) follows by the definition of $\Vdash^u$. The two implications (ii) $\Rightarrow$ (iii) and (iii) $\Rightarrow$ (iv) are part of Theorem 9.1. The implication (iv) $\Rightarrow$ (v) is trivial. Finally the implication (v) $\Rightarrow$ (i) is an immediate consequence of canonicity. ∎

In a sense this proof is a cheat since it has been achieved as a result of a judicious choice of notion (namely canonicity). To justify calling the result a theorem we must now exhibit a collection of interesting canonical systems.

10.2 Some canonical systems

So far the only canonical system we know is the smallest one K. In this section we give a couple more examples before, in the next section, we show how to generate a whole family of such systems.

Our first example is the system KD obtained by adding to the basic axioms all formulas

$$D(\phi) : \quad \Box\phi \to \Diamond\phi$$

(for arbitrary ϕ). We know this set of formulas characterizes seriality, so the following result is crucial.

10.4 LEMMA. *Let* S *be any system with* KD $\leq$ S. *Then the canonical structure* $\mathfrak{S}$ *of* S *is serial.*

Proof. Consider any $s \in S$ (where, of course, S is the carrier of $\mathfrak{S}$, i.e. the set of maximally S-consistent sets). We must produce some $t \in S$ with

$$s \longrightarrow t$$

i.e. such that for each formula ϕ

$$\Box\phi \in s \ \Rightarrow \ \phi \in t.$$

To this end let Φ be the set of all formulas ϕ for which

$$\Box \phi \in s.$$

It is sufficient to show that Φ is S-consistent (for then Lemma 9.4 provides some $t \in \boldsymbol{S}$ with $\Phi \subseteq t$, and hence $s \longrightarrow t$).

By way of contradiction suppose that Φ is not S-consistent. Then

$$\Phi \vdash_{\mathsf{S}}^w \bot$$

and hence there are $\phi_1, \ldots, \phi_n$ with

$$\vdash_{\mathsf{S}} \phi \to \bot \quad \text{i.e.} \quad \vdash_{\mathsf{S}} \neg \phi$$

where

$$\phi \quad \text{is} \quad \phi_1 \wedge \cdots \wedge \phi_n.$$

An application of (N) now gives

$$\vdash_{\mathsf{S}} \Box \neg \phi$$

hence, since $\mathsf{KD} \leq \mathsf{S}$, we may use $\mathrm{D}(\neg\phi)$ to get

$$\vdash_{\mathsf{S}} \Diamond \neg \phi.$$

Thus, rephrasing $\Diamond \neg \phi$, we have

$$\vdash_{\mathsf{S}} \neg \Box \phi \quad \text{i.e.} \quad \vdash_{\mathsf{S}} \Box \phi \to \bot.$$

Finally, since

$$\vdash_{\mathsf{K}} (\Box \phi_1 \wedge \cdots \wedge \Box \phi_n \to \Box \phi)$$

we see that $\Box \phi \in s$ and, more importantly,

$$s \vdash_{\mathsf{S}} \bot$$

which is the sought contradiction. ∎

Since the (unadorned) models of KD are precisely the serial structures, this result has an immediate consequence.

10.5 COROLLARY. *The system* KD *is canonical, and hence Kripke-complete.*

For our next example of canonicity consider the system KR formed by adding to the basic axioms all formulas

$$\mathrm{R}(\phi) \quad : \quad \Box^2 \phi \to \Box \phi$$

(for arbitrary ϕ). We know that the models of KR are precisely the dense structures, so our aim is to verify that the canonical structure of KR has this property. To do this we use what you may think is a rather obvious result.

10.6 LEMMA. *For an arbitrary formal system* S *let* $r, t \in S$ *be such that*

$$\Box^2 \phi \in r \;\Rightarrow\; \phi \in t$$

holds for all formulas ϕ. *Then we have*

$$r \longrightarrow s \longrightarrow t$$

for some $s \in S$.

Proof. Consider the set of formulas

$$\Phi = \{\theta \mid \Box\theta \in r\} \cup \{\Diamond\psi \mid \psi \in t\}.$$

We show that Φ is S-consistent.

By way of contradiction suppose otherwise, so that

$$\vdash_S (\theta_1 \wedge \cdots \wedge \theta_m \wedge \Diamond\psi_1 \wedge \cdots \wedge \Diamond\psi_n \to \bot)$$

for appropriate $\theta_1, \cdots, \theta_m, \psi_1, \cdots, \psi_n$. Let

$$\theta := \theta_1 \wedge \cdots \wedge \theta_m \quad \text{and} \quad \psi := \psi_1 \wedge \cdots \wedge \psi_n.$$

Note that

$$\vdash_K (\Box\theta_1 \wedge \cdots \wedge \Box\theta_m \to \Box\theta) \;,\quad \vdash_K (\Diamond\psi \to \Diamond\psi_1 \wedge \cdots \wedge \Diamond\psi_n)$$

so that $\Box\theta \in r$, and clearly $\psi \in t$. These implications also give

$$\vdash_S (\theta \wedge \Diamond\psi \to \bot)$$

which can be rephrased as

$$\vdash_S \theta \to \neg\Diamond\psi \quad \text{or} \quad \vdash_S \theta \to \Box\neg\psi$$

and hence an application of (EN) gives

$$\vdash_S \Box\theta \to \Box^2\neg\psi.$$

But now, since $\Box\theta \in r$, this gives $\Box^2\neg\psi \in r$, and hence $\neg\psi \in t$, which is the required contradiction.

This shows that Φ is S-consistent, and hence (by the Basic Existence Result) there is some $s \in S$ with $\Phi \subseteq s$, i.e. such that

$$\Box\theta \in r \;\Rightarrow\; \theta \in s \quad \text{and} \quad \psi \in t \;\Rightarrow\; \Diamond\psi \in s$$

for all formulas θ and ψ. The first of these gives $r \longrightarrow s$ and the second gives $s \longrightarrow t$, which is the desired result. ∎

Using this result we quickly obtain the analogue of Lemma 10.4.

10.7 LEMMA. *Let* S *be any system with* KR $\leq$ S. *Then the canonical structure* $\mathfrak{S}$ *of* S *is dense.*

Proof. Consider any $r, t \in S$ with $r \longrightarrow t$. This with the shape R shows that for each formula ϕ we have both

$$\square^2 \in r \;\Rightarrow\; \square \phi \in r \;\Rightarrow\; \phi \in t$$

and hence the required $s \in S$ with

$$r \longrightarrow s \longrightarrow t$$

is provided by Lemma 10.6. ∎

10.8 COROLLARY. *The system* KR *is canonical, and hence Kripke-complete.*

10.3 Confluence induced completeness

We have seen that the three systems

$$\mathsf{K} \;,\; \mathsf{KD} \;,\; \mathsf{KR}$$

are canonical, and hence Kripke-complete. Lemmas 10.4 and 10.7 also ensure that KDR is canonical. In fact these results are just particular instances of a quite widely applicable result making use of the confluence properties.

Recall from Chapter 6 that each 4-tuple of sequences of labels

$$
\begin{aligned}
\mathsf{i} &:= i(1), i(2), \ldots, i(p) \\
\mathsf{j} &:= j(1), j(2), \ldots, j(q) \\
\mathsf{k} &:= k(1), k(2), \ldots, k(r) \\
\mathsf{l} &:= l(1), l(2), \ldots, l(s)
\end{aligned}
$$

gives us two related gadgets. Firstly there is the structural property

$$CONF(\mathsf{i}; \mathsf{j}; \mathsf{k}; \mathsf{l})$$

which is concerned with the formation of a diagram of the following kind.

Secondly there is the set

$$Conf(\mathsf{i}; \mathsf{j}; \mathsf{k}; \mathsf{l})$$

of formulas

$$\langle i \rangle [j] \phi \;\to\; [k] \langle l \rangle \phi$$

(for arbitrary ϕ). We know that a structure has property $CONF(i; j; k; l)$ precisely when it models $Conf(i; j; k; l)$.

We now recognize $Conf(i; j; k; l)$ as an axiom schema; in particular we may form the system

$$K(i, j, k, l)$$

whose proper axioms are precisely the set $Conf(i; j; k; l)$. More generally, let Σ be any set of 4-tuples

$$\sigma = (i, j, k, l)$$

(of sequences of labels of varying length and components). Let

$$K(\Sigma)$$

be the system whose proper axioms are all the sets $Conf(\sigma)$ for $\sigma \in \Sigma$. For convenience let us refer to such a system as a *confluence system*. In this section we prove the following result.

10.9   THEOREM. *Each confluent system is canonical and consequently is also Kripke-complete.*

To prove this we use a generalization of Lemmas 10.4 and 10.7, and for this we need an extension of Lemma 10.6.

10.10   LEMMA. *Let S be an arbitrary formal system, let $r, t \in S$, and let*

$$i := i(1), i(2), \cdots i(p)$$

be a sequence of labels such that

$$[i]\phi \in r \;\Rightarrow\; \phi \in t$$

holds for all formulas ϕ. Then we have

$$r = s(0) \xrightarrow{\;i(1)\;} s(1) \xrightarrow{\;i(2)\;} \cdots \xrightarrow{\;i(p)\;} s(p) = t$$

for some $s(0), s(1), \cdots, s(p) \in S$.

Proof. This is proved by induction on the length p of i.

For the base case $p = 0$ we are given that

$$\phi \in r \;\Rightarrow\; \phi \in t$$

i.e. that $r \subseteq t$, and hence $r = t$. Thus we may set $s(0) = r = t$.

For the induction step $(p > 0)$ we may decompose i as the concatenation

$$i = j^\frown k$$

of two non-empty sequences $\mathbf{j}$ and $\mathbf{k}$ of labels. The property relating r and t is then

$$[\mathsf{j}]\,[\mathsf{k}]\phi \in r \;\Rightarrow\; \phi \in t$$

(for arbitrary ϕ). It is sufficient to produce some $s \in S$ such that

$$[\mathsf{j}]\phi \in r \;\Rightarrow\; \phi \in s \quad,\quad [\mathsf{k}]\phi \in s \;\Rightarrow\; \phi \in t$$

for then we may invoke the Induction Hypothesis to produce the required sequence.

The proof is now a replica of that of Lemma 10.6.

Thus consider the set of formulas

$$\Phi \;:=\; \{\theta \mid [\mathsf{j}]\theta \in r\} \cup \{\langle \mathsf{k}\rangle\psi \mid \psi \in t\}.$$

We show that Φ is S-consistent.

By way of contradiction, suppose otherwise, so that

$$\vdash_{\mathsf{S}} \theta_1 \wedge \cdots \wedge \theta_m \wedge \langle\mathsf{k}\rangle\psi_1 \wedge \cdots \wedge \langle\mathsf{k}\rangle\psi_n \;\to\; \bot$$

for appropriate $\theta_1, \cdots, \theta_m, \psi_1, \cdots, \psi_n$. Let

$$\theta := \theta_1 \wedge \cdots \wedge \theta_m \qquad \text{and} \qquad \psi := \psi_1 \wedge \cdots \wedge \psi_n.$$

Note that

$$\vdash_{\mathsf{K}} \big([\mathsf{j}]\theta_1 \wedge \cdots \wedge [\mathsf{j}]\theta_m \to [\mathsf{j}]\theta\big) \quad,\quad \vdash_{\mathsf{K}} \big(\langle\mathsf{k}\rangle\psi \to \langle\mathsf{k}\rangle\psi_1 \wedge \cdots \wedge \langle\mathsf{k}\rangle\psi_n\big)$$

so that $[\mathsf{j}]\theta \in r$, and clearly $\psi \in t$. These implications also give

$$\vdash_{\mathsf{S}} \theta \wedge \langle\mathsf{k}\rangle\psi \;\to\; \bot$$

which we can rephrase as

$$\vdash_{\mathsf{S}} \theta \;\to\; [\mathsf{k}]\neg\psi.$$

Several applications of (EN) now give

$$\vdash_{\mathsf{S}} [\mathsf{j}]\theta \;\to\; [\mathsf{j}]\,[\mathsf{k}]\neg\psi.$$

But now, since $[\mathsf{j}]\theta \in r$, this gives

$$[\mathsf{j}]\,[\mathsf{k}]\neg\psi \in r$$

which (by the given relationship between r and t) gives

$$\neg\psi \in t.$$

Since $\psi \in t$, this is the sought contradiction. ∎

We can now obtain the generalization of Lemmas 10.4 and 10.7.

10.11 LEMMA. *Let* i, j, k, l *be a fixed 4-tuple of sequences of labels, and let* S *be a formal system with* K(i, j, k, l) ≤ S. *Then the canonical structure* 𝔖 *of* S *enjoys* CONF (i; j; k; l).

Proof. Consider any $p, q, r \in \boldsymbol{S}$ with

holding. We must produce some $s \in \boldsymbol{S}$ with

holding. To this end consider the set of formulas

$$\Phi := \{\theta \mid [\text{j}]\theta \in q\} \cup \{\psi \mid [\text{l}]\psi \in r\}.$$

By Lemma 10.10 it suffices to show that Φ is S-consistent.

By way of contradiction, suppose that Φ is not S-consistent. Then (since [·] commutes with ∧) there are formulas θ and ψ with

$$[\text{j}]\theta \in q \quad , \quad [\text{l}]\psi \in r$$

and

$$\vdash_{\mathsf{S}} \theta \wedge \psi \rightarrow \bot.$$

This can be rephrased as

$$\vdash_{\mathsf{S}} \theta \rightarrow \neg\psi$$

and hence applications of (EN) give

$$\vdash_{\mathsf{S}} [\text{j}]\theta \rightarrow [\text{j}]\neg\psi$$

so that $[\text{j}]\neg\psi \in q$. But $p \xrightarrow{\text{i}} q$ so that $\langle\text{i}\rangle[\text{j}]\neg\psi \in p$ and hence, by $Conf(\text{i}; \text{j}; \text{k}; \text{l})$, we have $[\text{k}]\langle\text{l}\rangle\neg\psi \in p$. Finally, since

$$q \xrightarrow{\text{k}} r$$

we have $\langle\text{l}\rangle\neg\psi \in r$, i.e. $\neg[\text{l}]\psi \in r$ which is the required contradiction. ∎

Theorem 10.9 is now an immediate consequence of Lemma 10.11.

10.4 Exercises

10.1 Let S be any standard formal system all of whose axioms are sentences. Show that S is canonical. Hence show that KD is canonical.

10.2 Let E and F be the set of all formulas of the shapes

$$E : - \Diamond \top \to (\Box \phi \to \phi) \quad , \quad F : - \Box(\Box \phi \to \phi)$$

respectively (for arbitrary ϕ). Consider the two standard formal systems

$$S = KE \quad , \quad S = KF$$

separately.

(a) Find a correspondence result for S.

(b) Show that S is canonical.

10.3 For fixed label i, j, k, l, m, n let $K(i, j, k, l, m, n)$ be the formal system axiomatized by all the formulas $K(i, j, k, l, m, n)$ of Exercise 5.3 of Chapter 5. Show that this system is canonical.

10.4 Let S be the standard formal system axiomatized by all formulas of the shape

$$\Box(\Box \phi \to \psi) \vee \Box(\Box \psi \to \phi)$$

for arbitrary formulas ψ and ϕ. Recalling Proposition 5.5, show that S is canonical.

10.5 Let i, j, k, l, m, n be fixed labels and let S be the formal system axiomatized by the shape

$$(\langle i \rangle [l] \phi \wedge \langle k \rangle [n] \psi) \to [j] \langle m \rangle (\phi \wedge \psi)$$

(for arbitrary formulas ϕ, ψ). Show that S is canonical.

10.6 Let S be the formal system axiomatized by the formulas of the shape given in Exercise 6.3 of Chapter 6 (for the given parameters). Show that S is canonical.

10.7 Continuing with Exercises 8.12 and 8.13 let NATTIM be the extension of LINTIM formed by the addition of the axioms

- $[+]\top$

- $L_-(\phi)$

- $\langle + \rangle [+] \phi \to L_+(\phi)$

(for all formulas ϕ). Here L$_-$ and L$_+$ are the two Löb shapes formed using $[\text{-}]$ and $[\text{+}]$ respectively. Let $\mathcal{A}$ be an arbitrary model of NATTIM and set

$$I(\mathcal{A}) = \{a \in A \mid a \Vdash [\text{-}]\bot\}.$$

(a) Show that $\mathcal{N}$ models NATTIM but $\mathcal{Z}$ does not.

(b) Making use of descending chains and axiom L$_-$ show the following for each model $\mathcal{A}$ of NATTIM.

    (i) For each $b \in A$ there is a unique $a \in I(\mathcal{A})$ with $a \sim b$.

    (ii) There is a function

$$\text{next} : A \longrightarrow A$$

such that for each $a \in A$

$$a \xrightarrow{\ +\ } \text{next}(a)$$

and

$$a \xrightarrow{\ +\ } x \;\Rightarrow\; x = \text{next}(a) \text{ or } \text{next}(a) \xrightarrow{\ +\ } x$$

(for all $x \in A$).

    (iii) The function next is injective.

(c) Given a model $\mathcal{A}$ of NATTIM, some $a \in I(\mathcal{A})$, and some variable P, consider a valuation on $\mathcal{A}$ where

$$x \Vdash P \;\Leftrightarrow\; (\exists m \in \mathsf{N})[\text{next}^m(a) = x]$$

(for each $x \in A$). Making use of the axiom based on L$_+$ show the following.

    (i) $a \Vdash [\text{+}]\langle\text{+}\rangle P$.

    (ii) For each $b \in A$ there are unique $a \in I(\mathcal{A})$ and $m \in \mathsf{N}$ with $b = \text{next}^m(a)$.

    (iii) $\mathcal{A}$ consists of disjoint copies of $\mathcal{N}$.

(d) Show that

$$\mathcal{N} \Vdash^u \phi \;\Rightarrow\; \mathcal{A} \Vdash^u \phi$$

holds for all formulas ϕ.

(e) You are not asked to prove that NATTIM is Kripke-complete, however, assuming that this is true, show that

$$\vdash_{\text{NATTIM}} \phi \;\Leftrightarrow\; \mathcal{N} \Vdash^u \phi$$

holds for all formulas ϕ.

Part IV

Model constructions

This part begins the non basic part of the book and is built around a theme of semantics preserving morphisms. First in Chapter 11 the notion of a *bisimulation* is developed. This is a kind of relation between structures designed to highlight the similarities between them which, as a side product, has the required semantic preserving properties. The notion subsumes most other semantic preserving morphisms.

Building on this notion, in Chapter 12 a method of constructing smaller structures from larger structures is described. This method of *filtrations* has completeness and decidability consequences for the sytems to which it is applied. These are developed in Chapter 13.

Chapter 11

Bisimulations

11.1 Introduction

Both transition structures and valued transition structures are examples of relational structures. There is a standard way of comparing the algebraic properties of two such structures, namely through the notion of a *morphism*.

11.1 DEFINITION. (a) A morphism

$$\mathcal{A} \xrightarrow{\;\;f\;\;} \mathcal{B}$$

from a structure $\mathcal{A}$ to a structure $\mathcal{B}$ of the same signature is a function

$$f : A \longrightarrow B$$

from the carrier A of $\mathcal{A}$ to the carrier B of $\mathcal{B}$ such that

(Rel$^{\rightarrow}$) For each label i and elements a, x of $\mathcal{A}$

$$a \xrightarrow{\;i\;} x \quad \Rightarrow \quad f(a) \xrightarrow{\;i\;} f(x)$$

holds.

(b) A morphism

$$(\mathcal{A}, \alpha) \xrightarrow{\;\;f\;\;} (\mathcal{B}, \beta)$$

from a valued structure $(\mathcal{A}, \alpha)$ to a valued structure $(\mathcal{B}, \beta)$ is a morphism f between unadorned structures (as in (a)) such that

(Val$^{\rightarrow}$) For each variable P and element a of $\mathcal{A}$

$$a \in \alpha(P) \quad \Rightarrow \quad f(a) \in \beta(P)$$

holds. ∎

141

It is important to notice that these two defining conditions are *implications*, and not equivalences. All the standard universal algebraic constructions are carried out with these morphisms in mind, nevertheless there are situations where a more restricted notion of morphism is desirable. Modal logic is one of these cases.

Given a morphism f (as above) it is natural to consider which formulas are preserved or reflected by f, that is, the formulas ϕ such that for each element a of $\mathcal{A}$, one or other of the two implications

$$(\mathcal{A}, \alpha, a) \Vdash \phi \quad \Leftrightarrow \quad (\mathcal{B}, \beta, f(a)) \Vdash \phi \tag{11.1}$$

holds. For a general morphism f the corresponding class of formulas ϕ is not very interesting, hence the need to restrict the morphisms in some way.

Suppose first that we are interested in those morphisms f for which the equivalence (11.1) holds for all formulas ϕ and elements a of $\mathcal{A}$. A rather simple restriction on f will ensure this, and the resulting morphisms are variously known as *zigzag morphism* or *p-morphisms* (depending on the literary pretensions of the author). What is interesting about this restriction is that, once it has been elucidated, it becomes clear that it is also applicable to relations (as well as functions) between A and B. Those relations restricted in this way are called *bisimulations*, and this chapter is devoted to a full discussion of these.

A more extensive problem is to look for morphisms f for which the equivalence (11.1) holds for a restricted class of formulas. In this rather general form nothing beyond rather superficial results can be expected (for there are just too many parameters involved). There is, however, a particularly distinguished class of such morphisms known as *filtrations*. These are discussed in the next chapter.

11.2 Zigzag morphisms

In the literature you will find the terms 'zigzag morphism' and 'p-morphism' used more or less interchangeably. However, on closer inspection you will also find there are two closely related notions involved and it is instructive to make a pedantic distinction between these. Thus here we will use the two terms for the following notions

11.2 DEFINITION. We need two more kinds of morphisms.

(a) A *p-morphism*

$$\mathcal{A} \xrightarrow{\ \ f\ \ } \mathcal{B}$$

between two structures $\mathcal{A}$ and $\mathcal{B}$ of the same signature is a morphism f such that

(Rel$^{\leftarrow}$) For each label i and elements a of $\mathcal{A}$ and y of $\mathcal{B}$ with $f(a) \xrightarrow{i} y$, there is an element x of $\mathcal{A}$ with $a \xrightarrow{i} x$ and $f(x) = y$.

holds.

(b) A *zigzag morphism*

$$(\mathcal{A}, \alpha) \xrightarrow{\;f\;} (\mathcal{B}, \beta)$$

between two valued structures is a p-morphism f (as in (a)) such that

(Val$^{\leftarrow}$) For each variable P and element a of $\mathcal{A}$ the implication

$$a \in \alpha(P) \quad \Leftarrow \quad f(a) \in \beta(P)$$

holds. ∎

The condition (Val$^{\leftarrow}$) is the converse of the condition (Val$^{\rightarrow}$) of Section 11.1. Since f is a morphism it must also satisfy this condition so each zigzag morphism satisfies the equivalence

(Val$^{\leftrightarrow}$) $\qquad a \in \alpha(P) \quad \Leftrightarrow \quad f(a) \in \beta(P)$

(for all appropriate P and a). Note that given any p-morphism

$$\mathcal{A} \xrightarrow{\;f\;} \mathcal{B}$$

and valuation β on $\mathcal{B}$, the equivalence Val$^{\leftrightarrow}$ may be used to construct a valuation α on $\mathcal{A}$ such that f becomes a zigzag morphism

$$(\mathcal{A}, \alpha) \xrightarrow{\;f\;} (\mathcal{B}, \beta).$$

For this reason the distinction we have made between p-morphism and zigzag morphism need not be followed too enthusiastically.

The reason for introducing zigzag morphism is explained by the following result.

11.3 THEOREM. *Let*

$$(\mathcal{A}, \alpha) \xrightarrow{\;f\;} (\mathcal{B}, \beta)$$

be a zigzag morphism. Then

$$(\mathcal{A}, \alpha, a) \Vdash \phi \quad \Leftrightarrow \quad (\mathcal{B}, \beta, b) \Vdash \phi$$

holds for all elements a of $\mathcal{A}$ and formulas ϕ.

We need not prove this result here since it is a simple consequence of the broader analysis which is the main topic of this chapter.

11.3   Bisimulations

Given a pair

$$(\mathcal{A}, \alpha) \quad , \quad (\mathcal{B}, \beta)$$

of valued structures (of the same signature) we we interested in relations

$$R \subseteq A \times B$$

which connect the semantical properties of the structures. There are two such relations of particular interest.

11.4 DEFINITION. Consider two valued structures (as above).

($\sim$) Let $\sim$ be the relation between A and B defined by

$$a \sim b \quad \Leftrightarrow \quad (\forall P \in Var)[a \in \alpha(P) \Leftrightarrow b \in \beta(P)]$$

for $a \in A$ and $b \in B$. We say an arbitrary relation R is a *matching* if $R \subseteq \sim$.

($\approx$) Let $\approx$ be the relation between A and B such that for all $a \in \mathcal{A}$ and $b \in \mathcal{B}$

$$a \approx b$$

holds precisely when the equivalence

$$(\mathcal{A}, \alpha, a) \Vdash \phi \quad \Leftrightarrow \quad (\mathcal{B}, \beta, b) \Vdash \phi$$

holds for all formulas ϕ. We call $\approx$ the *semantic equivalence* relation. ■

Note that the semantic equivalence relation $\approx$ is a matching. Our objective is to look for various approximations to $\approx$.

11.5 DEFINITION. Consider two valued structures (as above).

(a) A relation R has the *back and forth* property if for each label i and elements a of $\mathcal{A}$ and b of $\mathcal{B}$, if aRb then both

back $(\forall y \prec_i b)(\exists x \prec_i a)[xRy]$
forth $(\forall x \prec_i a)(\exists y \prec_i b)[xRy]$

hold.

(b) A *bisimulation* is a matching which has the back and forth property. ■

For example, the empty relation is a bisimulation (vacuously), however we will see more interesting examples later. Before we do that let us see how $\approx$ and bisimulations are connected.

11.6 THEOREM. *Each bisimulation between two valued structures $(\mathcal{A}, \alpha)$ and $(\mathcal{B}, \beta)$ is included in the semantic equivalence relation.*

Proof. Let R be the given bisimulation. For each formula ϕ consider the following condition

(ϕ)   For each $a \in A$ and $b \in B$, if aRb then, $a \Vdash \phi \Leftrightarrow b \Vdash \phi$.

We show that (ϕ) holds for all formulas ϕ by induction on the complexity of ϕ.

The base case holds since R is a matching, and the passage across the boolean connectives is immediate. Thus it suffices to consider the passage across $[i]$ for an arbitrary label i.

Consider any formula $[i]\phi$ and elements a and b with aRb . Then

$$a \Vdash [i]\phi \Leftrightarrow (\forall x \prec_i a)[x \Vdash \phi] \Leftrightarrow (\forall y \prec_i b)[y \Vdash \phi] \Leftrightarrow b \Vdash [i]\phi$$

where the central equivalence is verified using the Induction Hypothesis and the back and forth property, as follows.

Suppose $x \Vdash \phi$ for each $x \prec_i a$ and consider any $y \prec_i b$. The back condition gives us a particular $x \prec_i a$ with xRy. But then (ϕ) gives

$$x \Vdash \phi \quad \Leftrightarrow \quad y \Vdash \phi$$

and hence $y \Vdash \phi$. This proves one half of the equivalence and the converse follows in a similar fashion.

This completes all the required induction steps. ∎

This result shows that the semantic equivalence relation $\approx$ sits below $\sim$ and above all bisimulations. We need to close this gap, but before we do that let us see how bisimulations subsume zigzag morphisms.

Each morphism

$$\mathcal{A} \xrightarrow{\quad f \quad} \mathcal{B}$$

gives us a relation $F \subset A \times B$, namely its graph defined by

$$aFb \quad \Leftrightarrow \quad f(a) = b$$

(for $a \in A$ and $b \in B$).

11.7 THEOREM. *Consider any morphism f with graph F, as above.*

(a) *This morphism f is a p-morphism precisely when F has the back and forth property*

(b) *Given valuations on the structures, the morphism f is a zigzag morphism precisely when the F is a bisimulation.*

Proof. We will actually prove something a bit more detailed. Thus consider any function

$$f : A \longrightarrow B.$$

We show the following.

(i) The function f has Rel$^{\rightarrow}$ (and hence is a morphism) precisely when F has the forth property.

(ii) The function f has Rel$^{\leftarrow}$ precisely when F has the back property.

(iii) Given valuations α on $\mathcal{A}$ and β on $\mathcal{B}$, the function f has Val$^{\leftrightarrow}$ precisely when F is a matching.

We prove these three correspondences in turn.

(i) Suppose first that f has Rel$^{\rightarrow}$ and consider any $a, x \in A$ and $b \in B$ with

$$aFb \quad , \quad a \xrightarrow{i} x.$$

Then $f(a) = b$ so that, by Rel$^{\rightarrow}$, we have $b \xrightarrow{i} f(x)$, and hence we may set $y = f(x)$ to verify the forth property. Conversely, suppose that F has the forth property, and consider any $a, x \in A$ with $a \xrightarrow{i} x$. Setting $b = f(x)$ we have aFb so that (by the forth property) there is some $y \in B$ with

$$xFy \quad , \quad b \xrightarrow{i} y.$$

Since $y = f(x)$, this gives $f(a) \xrightarrow{i} f(x)$, as required.

(ii) Suppose first that f has Rel$^{\leftarrow}$ and consider any $a \in A$ and $b, y \in B$ with

$$aFb \quad , \quad b \xrightarrow{i} y.$$

Then $f(a) = b$ so that $f(a) \xrightarrow{i} y$ and hence, by Rel$^{\leftarrow}$ there is some $x \in A$ with

$$f(x) = y \quad , \quad a \xrightarrow{i} x.$$

This verifies the back property. The converse implication follows in the same way.

(iii) This is immediate. ∎

This result when combined with Theorem 11.6 gives a proof of Theorem 11.3.

We will say nothing more about zigzag morphisms; all we need to know about them can be deduced as the functional case of a bisimulation.

11.4   The largest bisimulation

For the remainder of the chapter let

$$(\mathcal{A}, \alpha) \quad , \quad (\mathcal{B}, \beta)$$

be a fixed pair of valued structures. We know there is at least one bisimulation between this pair, namely the empty relation. This, however, does not hold the attention for very long (even though it may be the only bisimulation). A more interesting example is at the opposite end of the scale.

11.8 THEOREM. *There is a unique largest bisimulation i.e. a bisimulation which includes all other bisimulations.*

Proof. Let $\mathcal{R}$ be the family of all bisimulations. We know that $\mathcal{R}$ is non-empty (since $\emptyset \in \mathcal{R}$). Set

$$S = \bigcup \mathcal{R}$$

i.e. let S be the relation such that

$$aSb$$

holds (for $a \in A$ and $b \in B$) precisely when there is some $R \in \mathcal{R}$ with aRb. Clearly S is a matching (and, in fact, it is included in $\approx$) and includes all bisimulations. Thus it suffices to show that S has the back and forth property.

For a fixed label i consider any $a \in A$ and $b \in B$. Consider also any $x \in A$ with $x \prec_i a$. By definition of S there is some $R \in \mathcal{R}$ with aRb. But then, since this R is a bisimulation, there is some $y \prec_i b$ with xRy. In particular, xSy, and so we have verified the forth property. The back property is verified in the same way. ∎

Because of its special position we let

$$\approx$$

be this largest bisimulation. In particular we have

$$\approx \; \subseteq \; \approx \; \subseteq \; \sim$$

which gives us lower and upper bounds for $\approx$. In general these three relations are distinct, but there is an interesting situation where the two lower ones agree.

We say a structure $\mathcal{A}$ is *image finite* if for each label i and element a of $\mathcal{A}$, there are just finitely many elements x with

$$a \xrightarrow{\; i \;} x.$$

In particular, every finite structure and every deterministic structure is image finite.

11.9 THEOREM. *Suppose both the structures $\mathcal{A}$ and $\mathcal{B}$ are image finite. Then the two relations $\cong$ and $\approx$ coincide.*

Proof. It suffices to show that $\approx$ has the back and forth property. We will verify the forth property; the verification of the back property then follows in a similar fashion.

Consider elements $x, a \in A$ and $b \in B$ with $a \approx b$ and $x \prec_i a$ (for some label i). We must produce some $y \in B$ with $y \prec_i b$ and $x \approx y$.

Let $y_1, \ldots, y_n$ be all the elements $y \in B$ with $y \prec_i b$. We may restrict our search to this finite set. By way of contradiction suppose there is no such y_r with $x \approx y_r$. Then for each $1 \leq r \leq n$, there is a formula θ_r with

$$x \Vdash \neg\theta_r \quad , \quad y_r \Vdash \theta_r.$$

Let ϕ be

$$\theta_1 \vee \ldots \vee \theta_n$$

so that $y_r \Vdash \phi$ for each $1 \leq r \leq n$, and hence

$$b \Vdash \square \phi.$$

But then, since $a \approx b$, we have

$$a \Vdash \square \phi$$

and hence $x \Vdash \phi$ which produces some $1 \leq r \leq n$ with

$$x \Vdash \theta_r.$$

Since this is a direct contradiction, the proof is complete. ∎

11.5   A hierarchy of matchings

The construction of $\cong$ given in the proof of Theorem 11.8 does not provide much information beyond that $\cong$ is a bisimulation in a special position. There is another construction of $\cong$ (this time from above rather than below) which also provides a measure of its complexity. To describe this we need some preliminaries.

Given a relation R, its *derivative* is the relation $R^{\blacktriangledown}$ defined such that, for each $a \in A$ and $b \in B$,

$$aR^{\blacktriangledown}b$$

holds precisely when

$$aRb$$

and for each label i both

- $(\forall x \prec a)(\exists y \prec_i b)[xRy]$

- $(\forall y \prec b)(\exists x \prec_i a)[xRy]$

hold. Trivially $R^{\blacktriangledown} \subseteq R$. Note also that the operation $(\cdot)^{\blacktriangledown}$ is monotone, i.e.

$$S \subseteq R \quad \Rightarrow \quad S^{\blacktriangledown} \subseteq R^{\blacktriangledown}$$

holds for all relations R and S.

Bisimulations are precisely those matchings which are the fixed points of $(\cdot)^{\blacktriangledown}$, i.e. matchings R such that $R^{\blacktriangledown} = R$. This enables us to apply a standard procedure for obtaining fixed points.

Let Ord be the class of ordinals. For a given relation R we define the descending chain

$$(R_\alpha \mid \alpha \in Ord)$$

by an iteration of $(\cdot)^{\blacktriangledown}$. Thus we set

$$R_0 = R \quad , \quad R_{\alpha+1} = (R_\alpha)^{\blacktriangledown} \quad , \quad R_\lambda = \bigcap \{R_\alpha \mid \alpha < \lambda\}$$

for each ordinal α and limit ordinal λ. On cardinality grounds this chain eventually stabilizes, i.e. there is some ordinal ∞ such that

$$R_\alpha = R_\infty$$

for all ordinals $\alpha \geq \infty$. In fact, ∞ is the first ordinal α such that $R_{\alpha+1} = R_\alpha$. In particular, if the parent relation R is a matching, then the stable descendant R_∞ is a bisimulation.

11.10 THEOREM. *For each matching R, the matching R_∞ is the largest bisimulation included in R.*

Proof. We have already noted that R_∞ is a bisimulation and $R_\infty \subseteq R$. Consider any other bisimulation $S \subseteq R$. Then, using the monotone property of $(\cdot)^{\blacktriangledown}$ we have

$$S = S^{\blacktriangledown} \subseteq R^{\blacktriangledown} = R_1.$$

In the same way an obvious induction shows that

$$S \subseteq R_\alpha$$

for all $\alpha \in Ord$; in particular, $S \subseteq R_\infty$, as required. ∎

Since $\approx$ is the largest bisimulation and is included in $\sim$, it is the largest bisimulation included in $\sim$. Thus we may apply the above construction to obtain $\approx$ from $\sim$. We let

$$\sim_0 \; = \; \sim \quad , \quad \sim_{\alpha+1} \; = \; (\sim_\alpha)^{\blacktriangledown} \quad , \quad \sim_\lambda \; = \; \bigcap \{\sim_\alpha \mid \alpha < \lambda\}$$

for each ordinal α and limit ordinal λ. Then $\approx$ is just $\sim_\infty$. This ordinal ∞ is a measure of the distance between $\sim$ and $\approx$, and hence tells us something about the complexity of $\approx$.

11.6   An example

At this stage in the proceedings it is instructive to look at a class of (particularly simple) examples. These examples show that the required value of ∞ can be indefinitely large.

In all the examples the two valued structures $(\mathcal{A}, \alpha)$ and $(\mathcal{B}, \beta)$ are the same, so we may concentrate on just one of them, $\mathcal{A}$. Also, for each variable P, we set

$$\alpha(P) = A.$$

This means that the relation $\sim$ is just $A \times A$, i.e. that $a \sim b$ holds for all $a, b \in A$.

The structure $\mathcal{A}$ is monomodal, in fact

$$\mathcal{A} = (A, \longrightarrow)$$

where A is an ordinal and, for $a, b \in A$

$$a \longrightarrow b \quad \Leftrightarrow \quad b < a$$

(where $<$ is the standard ordering on A).

For these examples we have a concise description of the relations $\sim_\alpha$.

11.11   PROPOSITION. *For each ordinal α and $a, b \in A$, the conditions*

(i) $a \sim_\alpha b$
(ii) $a = b$ *or* $\alpha \le a, b$

are equivalent.

Proof. This is proved by induction on α.

The base case $\alpha = 0$ is trivial since $a \sim_0 b$ and $0 \le a, b$ hold for all a, b.

For the induction step $\alpha \mapsto \alpha + 1$, suppose first that $a \sim_{\alpha+1} b$ and that $a \ne b$. Then $a < b$ (say) so, setting $y = a$ the back property of $\sim_{\alpha+1} = (\sim_\alpha)^\blacktriangledown$ produces some $x < a$ with $x \sim_\alpha y$. Since $x \ne y$ the Induction Hypothesis gives $\alpha \le x, a$, so that

$$\alpha \le x < a < b$$

and hence $\alpha + 1 \le a, b$.

Conversely, suppose that $\alpha + 1 \le a, b$ and consider any $x < a$, We require some y, b with $x \sim_\alpha y$. But if $x > a$ then we may take $y = x$, and if $\alpha \le x$ then we may take $y = \alpha$, for in both cases the Induction Hypothesis gives $x \sim_\alpha y$.

For the induction leap to a limit ordinal λ we argue

$$
\begin{aligned}
a \sim_\lambda b \quad &\Leftrightarrow \quad (\forall \alpha < \lambda)[a \sim_\alpha b] \\
&\Leftrightarrow \quad (\forall \alpha < \lambda)[a = b \text{ or } \alpha \le a, b] \\
&\Leftrightarrow \quad a = b \text{ or } (\forall \alpha < \lambda)[\alpha \le a, b] \quad \Leftrightarrow \quad a = b \text{ or } \lambda \le a, b
\end{aligned}
$$

which is the required result. ■

This class of examples shows that for each ordinal α the two relations $\sim_\alpha$ and $\sim_{\alpha+1}$ can be distinct. For let A be an ordinal which is at least $\alpha + 2$. Then

$$a = \alpha \quad , \quad b = \alpha + 1$$

are both members of A, and clearly

$$a \sim_\alpha b \quad , \quad \text{not}[a \sim_{\alpha+1} b].$$

which witnesses the required distinctness.

11.7 Stratified semantic equivalence

We now return to considering the fixed pair

$$(\mathcal{A}, \alpha) \quad , \quad (\mathcal{B}, \beta)$$

of valued structures. Each set of formulas Γ gives us a relation $|\,\Gamma\,|$ between these structures where, by definition,

$$a \mid \Gamma \mid b$$

holds (for $a \in A$ and $b \in B$) precisely when

$$(\forall \phi \in \Gamma)[a \Vdash \phi \quad \Leftrightarrow \quad b \Vdash \phi].$$

Thus, for example,

$$|\, Var \,| \text{ is } \sim \quad \text{ and } \quad |\, Form \,| \text{ is } \approx.$$

Notice that many different sets Γ can give the same relation $|\,\Gamma\,|$. In particular, if Γ^B is the boolean closure of Γ, then

$$|\, \Gamma^B \,| \quad = \quad |\, \Gamma \,|.$$

These relations $|\,\Gamma\,|$ are most useful when Γ is closed under subformulas, but we need not assume this in general.

For each set Γ let $\Gamma^\square$ be the set of formulas of the form

$$\phi \quad \text{or} \quad [i]\phi$$

for some $\phi \in \Gamma$ and label i. We are interested in the comparison between $|\,\Gamma\,|$ and $|\,\Gamma^\square\,|$.

11.12 LEMMA. *For each set of formulas Γ, we have $|\,\Gamma\,|^{\blacktriangledown} \subseteq |\,\Gamma^\square\,|$.*

Proof. The argument for this is essentially the same as the argument verifying the induction step in the proof of Theorem 11.6. ∎

Next let

$$\Delta_0 = Var \cup \{\top, \bot\}$$

so that $(\Delta_0)^B$ is the set of Box-free (i.e. propositional) formulas and hence (using the above remark)

$$| \Delta_0 | \; = \; \sim \; = \; \sim_0 .$$

Now for each $r < \omega$ set

$$\Delta_{r+1} = (\Delta_r^B)^\Box$$

to produce an ascending chain

$$\Delta_0 \subseteq \Delta_0^B \subseteq \cdots \subseteq \Delta_r \subseteq \Delta_r^B \subseteq \Delta_{r+1} \subseteq \cdots$$

with

$$\Delta_\omega = \bigcup \{\Delta_r \mid r < \omega\} = Form$$

and hence

$$| \Delta_\omega | \; = \; \approx .$$

We are interested in the intermediate relations $| \Delta_r |$.

11.13 LEMMA. *For each $r < \omega$, we have* $\sim_r \; \subseteq \; | \Delta_r |$.

Proof. We have seen already that

$$\sim_0 \; = \; | \Delta_0 | .$$

Now assuming the inclusion holds for r, we have

$$\sim_{r+1} \; = \; \sim_r^{\blacktriangledown} \; \subseteq \; | \Delta_r |^{\blacktriangledown} \; = \; | \Delta_r^B |^{\blacktriangledown} \; \subseteq \; | \Delta_r^{B\Box} | \; = \; | \Delta_{r+1} |$$

where the crucial step (the second inclusion) follows by Lemma 11.12.

The required result now follows by induction. ∎

Taking the limiting case of this gives the following.

11.14 COROLLARY. $\sim_\omega \; \subseteq \; \approx.$

We have seen already a condition which ensures that the two relations $\sim_\omega$ and $\approx$ coincide, namely image finiteness. But then $\approx$ also coincides with $\approx$. It is, therefore, of interest to find natural conditions which force $\sim_\omega$ and $\approx$ to agree but where these relations may still be distinct from $\approx$.

To this end, let us say a set of formulas Γ is *essentially finite* for the pair of valued structures in question, if there are finitely many members $\gamma_1, \ldots, \gamma_n$ of Γ such that for each $\phi \in \Gamma$ there is such a γ with

$$(\mathcal{A}, \alpha), (\mathcal{B}, \beta) \Vdash^v (\phi \leftrightarrow \gamma).$$

For example, any finite set is essentially finite. Also the set *Var* is essentially finite for the ordinal example of Section 11.6.

Recall that we say the signature is finite if there are just finitely many labels.

11.15 LEMMA. *Suppose the signature is finite and let Γ be an essentially finite set of formulas. Then Γ^B and $\Gamma^\square$ are also essentially finite.*

Proof. Let $\gamma_1, \ldots, \gamma_n$ be the given finitely many members of Γ which span Γ. We are interested in the formulas ψ of the form

$$\pm\gamma_1 \wedge \ldots \wedge \pm\gamma_n$$

where each $\pm\gamma$ is either γ or $\neg\gamma$. Notice that there are just finitely many such formulas ψ.

Each boolean combination of $\gamma_1, \ldots, \gamma_n$ is tautologically equivalent to a formula ϕ of the shape

$$\psi_1 \vee \ldots \vee \psi_m$$

where each formula ψ_r is a formula of the kind ψ just described. Thus we see that Γ^B is essentially finite.

Finally, since the set of labels is finite, we see that $\Gamma^\square$ is essentially finite. ∎

The next result explains why we have introduced this notion of essential finiteness.

11.16 PROPOSITION. *If the set of formulas Γ is essentially finite then the two relations $\mid \Gamma \mid^{\mathbf{v}}$ and $(\Gamma^\square)$ coincide.*

Proof. By Lemma 11.12 we know already that

$$\mid \Gamma \mid^{\mathbf{v}} \subseteq (\Gamma^\square).$$

Thus it suffices to show the converse inclusion, assuming that Γ is essentially finite.

Let $\gamma_1, \ldots, \gamma_n$ be the given spanning members of Γ. Consider any $a \in A$ and $b \in B$ with $a \mid \Gamma^\square \mid b$. Consider also any $x \prec_i a$ (for some label i). We must produce some $y \prec_i b$ with $x \mid \Gamma \mid y$.

Set

$$\psi \quad := \quad \pm\gamma_1 \wedge \ldots \wedge \pm\gamma_n$$

where each $\pm\gamma$ is γ or $\neg\gamma$ with the sign chosen so that

$$x \Vdash \pm\gamma.$$

Note that

$$x \Vdash \psi \quad , \quad \psi \in \Gamma^\square \quad , \quad [i]\neg\psi \in \Gamma^\square.$$

Now $a \Vdash \langle i \rangle\psi$ so that $b \Vdash \langle i \rangle\psi$ (since $a \mid \Gamma^\square \mid b$) which gives some $y \prec_i b$ with $y \Vdash \psi$. It suffices to show that $x \mid \Gamma \mid y$. But for each $\phi \in \Gamma$ there is some γ with

$$x \Vdash (\phi \leftrightarrow \gamma) \quad , \quad y \Vdash (\phi \leftrightarrow \gamma)$$

and hence

$$x \Vdash \phi \quad \Leftrightarrow \quad y \Vdash \phi$$

by the relationship of x and y to ψ. ■

Finally we have arrived at the result we have been travelling towards.

11.17 THEOREM. *Suppose that the signature is finite and that* Var *is essentially finite. Then for each* $r < \omega$ *the two relations* $\sim_r$ *and* $\mid \Delta_r \mid$ *agree. In particular* $\sim_\omega$ *and* $\approx$ *coincide.*

Proof. By definition, $\sim_0$ and $\mid \Delta_0 \mid$ agree. Also, assuming that $\sim_r$ and $\mid \Delta_r \mid$ agree, an application of Proposition 11.16 gives

$$\sim^{r+1} \;=\; (\sim_r)^{\blacktriangledown} \;=\; (\mid \Delta_r \mid)^{\blacktriangledown} \;=\; \mid (\Delta_r)^\square \mid \;=\; (\mid \Delta_{r+1} \mid)$$

so the required result follows by induction. ■

This result shows that for the ordinal examples of Section 11.6 the two relations $\sim_\omega$ and $\approx$ coincide. Notice also that for these examples the relation $\approx$ can be much smaller.

11.8   Exercises

11.1 Structures of the various kinds form several categories.

(a) Show that the composite of two morphisms (between structures of the same kind) is itself a morphism.

(b) Show that the composite of two p-morphisms is itself a p-morphism, and the composite of two zigzag morphisms is a zigzag morphism.

11.2 For the monomodal language, the singleton set $\{0\}$ carries just two transition structures

$$\mathcal{L} \quad , \quad \mathcal{R}$$

where the transition relation on $\mathcal{L}$ is empty and on $\mathcal{R}$ it is not.

segment

(a) Show that the unique assignment $g : A \longrightarrow \{0\}$ defines a morphism

$$A \xrightarrow{g} \mathcal{R}$$

and that this morphism is a p-morphism precisely when $\mathcal{A}$ is serial.

(b) Show that the morphisms

$$\mathcal{L} \xrightarrow{f} \mathcal{A}$$

are in bijective correspondence with the elements of $\mathcal{A}$, and that the p-morphisms (in this direction) are in bijective correspondence with the blind elements of $\mathcal{A}$.

(c) Determine the morphisms

$$A \longrightarrow \mathcal{L} \quad , \quad \mathcal{R} \longrightarrow A.$$

Which of these are p-morphisms?

11.3 Let

$$A \xrightarrow{f} B$$

be a p-morphism and let α be a valuation on $\mathcal{A}$. Show there are two valuations λ and ρ on $\mathcal{B}$ such that the following hold.

(a) Both

$$(\mathcal{A}, \alpha) \xrightarrow{f} (\mathcal{B}, \lambda) \quad , \quad (\mathcal{A}, \alpha) \xrightarrow{f} (\mathcal{B}, \rho)$$

are zigzag morphisms;

(b) For each valuation β on $\mathcal{B}$,

$$(\mathcal{A}, \alpha) \xrightarrow{f} (\mathcal{B}, \beta)$$

is a zigzag morphism if and only if $\lambda \leq \beta \leq \rho$, i.e.

$$\lambda(P) \subseteq \beta(P) \subseteq \rho(P)$$

holds for all variables P.

11.4 Let $\mathcal{A}$ and $\mathcal{B}$ be structures with $\mathcal{A} \subseteq \mathcal{B}$, and let f be the insertion $f : A \longrightarrow B$.

(a) Show that f is a morphism.

(b) Show that $\mathcal{A} \subseteq_g \mathcal{B}$ (in the sense of Exercise 4.9) if and only f is a p-morphism.

11.5 Let
$$R \subseteq A \times B \quad , \quad S \subseteq B \times C$$
be two relations. The sequential composite
$$R;S \subseteq A \times C$$
is the relation defined by
$$a(R;S)c \iff (\exists b \in B)[aRbSc]$$
(for $a \in A$ and $c \in C$).

(a) Show that the sequential composite of two back and forth relations is a back and forth relation, and the sequential composite of two bisimulations is a bisimulation.

(b) Let
$$f : A \longrightarrow B \quad , \quad g : B \longrightarrow C$$
be two functions with graphs F and G. Show that the sequential composite $F;G$ is the graph of the function composite gf.

11.6 Consider the structure
$$\mathcal{N} = (\mathsf{N}, \longrightarrow)$$
carried by the set of natural numbers N with the successor relation given by
$$a \longrightarrow b \iff a = b + 1$$
(for $a, b \in \mathsf{N}$). Let ν be the valuation on $\mathcal{N}$ with $\nu(P) = \mathsf{N}$ for all variables P, and consider the induced matching hierarchy $\sim_\bullet$.

(a) Show that for each $a, b \in \mathsf{N}$ and $r < \omega$
$$a \sim_{r+1} b \iff \begin{cases} a = b \leq r \\ \text{or} \\ a, b > r. \end{cases}$$

Hence show that $\sim_\omega = \sim_\infty$ is just equality.

(b) For each $k \in \mathsf{N}$, find a sentence ϕ_k such that
$$a \Vdash \phi_k \iff a = k$$
holds for all $a \in \mathsf{N}$.

Chapter 12

Filtrations

12.1 Introduction

In Chapter 11 we isolated a class of valued morphisms

$$(\mathcal{A}, \alpha) \longrightarrow (\mathcal{B}, \beta)$$

namely the zigzag morphisms, for which the equivalence

$$(\mathcal{A}, \alpha, a) \Vdash \phi \quad \Leftrightarrow \quad (\mathcal{B}, \beta, f(a)) \Vdash \phi$$

holds for all elements $a \in A$ and *all* formulas ϕ. These morphisms are most useful when the two structures are are given independently and when we really are concerned with all formulas. However, there are many situations where we are given only the valued structure $(\mathcal{A}, \alpha)$ and we are required to construct a valued structure $(\mathcal{B}, \beta)$ together with an appropriate morphism f such that the equivalence holds only for a restricted class of formulas. More often than not we also require $\mathcal{B}$ to be finite and as small as possible. In this chapter we will look at the commonest method of obtaining such morphisms, namely the method of filtrations. As we will see later in Chapter 13 this method has some significant consequences for the completeness and decidability of various formal systems.

Throughout the chapter Γ is some fixed set of formulas which is assumed to be closed under subformulas. In most applications this set Γ is the set of all subformulas of a given formula. Our objective is to ensure that the equivalence holds for all formulas $\phi \in \Gamma$.

12.1 DEFINITION. Let $(\mathcal{A}, \alpha)$ and $(\mathcal{B}, \beta)$ be two valued structures.

(a) A Γ-*morphism* from the first to the second is a morphism

$$\mathcal{A} \xrightarrow{\ f\ } \mathcal{B}$$

157

such that for each variable $P \in \Gamma$

$$a \in \alpha(P) \quad \Rightarrow \quad f(a) \in \beta(P)$$

holds for all $a \in A$.

(b) A Γ-*filtration* from the first structure to the second structure is a Γ-morphism f, as above, such that:

(Sur) This morphism is surjective.

(Var) For each variable $P \in \Gamma$ and element $a \in A$

$$a \in \alpha(P) \quad \Leftrightarrow \quad f(a) \in \beta(P).$$

(Fil) For each label i and formula ϕ with $[i]\phi \in \Gamma$, the implication

$$a \Vdash [i]\phi \quad \Rightarrow \quad x \Vdash \phi$$

holds for all elements a, x of $\mathcal{A}$ with $f(a) \xrightarrow{\ i\ } f(x)$ in $\mathcal{B}$. ∎

Notice that there are no restrictions at all on the variables $P \notin \Gamma$, in particular the filtration f need not be a full valued morphism.

The first thing to do now is to demonstrate why filtrations are useful.

12.2 THEOREM. *Let f be a Γ-filtration (as above). Then, for each formula $\phi \in \Gamma$ the equivalence*

$$(\mathcal{A}, \alpha, a) \Vdash \phi \quad \Leftrightarrow \quad (\mathcal{B}, \beta, f(a)) \Vdash \phi$$

holds for all elements a of $\mathcal{A}$.

Proof. We proceed by induction on ϕ.

The two base cases $\phi = \top$ and $\phi = \bot$ hold trivially. For the case $\phi = P \in Var$; either $P \in \Gamma$ in which case the equivalence is given by (Var), or $P \notin \Gamma$ in which case there is nothing to prove. The induction steps across propositional connectives follow easily since Γ is closed under subformulas, thus it remains to deal with the induction step across a box.

Fix a label i and consider any formula ϕ with $[i]\phi \in \Gamma$. Since we also have $\phi \in \Gamma$ the Induction Hypothesis gives

$$x \Vdash \phi \quad \Leftrightarrow \quad f(x) \Vdash \phi$$

for all $x \in A$. We must show that

$$a \Vdash [i]\phi \quad \Leftrightarrow \quad f(a) \Vdash [i]\phi$$

for all $a \in A$.

Suppose first that $a \Vdash [i]\phi$ and consider any $y \in B$ with $f(a) \xrightarrow{i} y$. Since f is surjective there is some $x \in A$ with $f(x) = y$. But then (Fil) gives $x \Vdash \phi$ and the Induction Hypothesis gives $y \Vdash \phi$, so that $f(a) \Vdash [i]\phi$. Conversely, suppose that $f(a) \Vdash [i]\phi$ and consider any $x \in A$ with $a \xrightarrow{i} x$. Since f is a morphism we have $f(a) \xrightarrow{i} f(x)$, so that $f(x) \Vdash \phi$ and hence the Induction Hypothesis gives $x \Vdash \phi$. Thus $a \Vdash [i]\phi$ as required. ∎

In this proof I have been a little bit sloppy because I did not state precisely what the Induction Hypothesis is. Before you continue you should do this and make sure you understand the mechanism of the induction.

12.2 The canonical carrying set

We have already fixed the set Γ. We now fix the valued structure $(\mathcal{A}, \alpha)$ and turn to the central topic of this chapter.

For the given $(\mathcal{A}, \alpha)$ and Γ, how can we construct a Γ-filtration of $(\mathcal{A}, \alpha)$ for which the target is as small as possible ?

Consider the equivalence relation $\sim$ on A (the carrying set of $\mathcal{A}$) defined by

$$a \sim x \quad \Leftrightarrow \quad (\forall \phi \in \Gamma)[a \Vdash \phi \Leftrightarrow x \Vdash \phi]$$

for all $a, x \in A$ and set

$$B = A/\sim$$

i.e. let B be the set of equivalence classes of A. For each $a \in A$ let $a^\sim$ be the equivalence class to which a belongs, i.e. let

$$a^\sim = \{x \in A \mid a \sim x\}.$$

Let

$$A \xrightarrow{f} B$$
$$a \longmapsto a^\sim$$

be the canonical surjection. We wish to construct a valued structure $(\mathcal{B}, \beta)$ on B in such a way that f becomes a Γ-filtration.

Before we do this let us estimate the size of B.

12.3 LEMMA. *Suppose the set of formulas Γ is finite. Then so is B and, in fact, $card(B) \leq 2^{card(\Gamma)}$.*

Proof. The set

$$[\Gamma \to 2]$$

of all functions from Γ to 2 is finite with cardinality $2^{card(\Gamma)}$. For each $a \in \Gamma$ let

$$f_a : \Gamma \longrightarrow 2$$

be the function given by

$$f_a(\phi) \;=\; \begin{cases} 1 \text{ if } a \Vdash \phi \\[2mm] 0 \text{ if } a \Vdash \neg\phi \end{cases}$$

(for $\phi \in \Gamma$). The definition of $\sim$ can be rephrased as

$$a \sim x \quad \Leftrightarrow \quad f_a = f_x$$

in particular we have a well-defined injection

$$B \longrightarrow [\Gamma \to 2]$$
$$a^{\sim} \longmapsto f_a$$

which gives the required result. ∎

Our main problem is to convert B into a transition structure satisfying the appropriate conditions. It turns out that there are many ways of doing this, but there is a 'left-most' (or $\exists$-) solution and a 'right-most' (or $\forall$-) solution. We look at these two cases in detail.

12.3   The left-most filtration

For each label i let $\overset{i}{\longrightarrow}_l$ be the relation on $B = A/\!\sim$ given by

$$b \overset{i}{\longrightarrow}_l y \quad \Leftrightarrow \quad (\exists a \in b, x \in y)[a \overset{i}{\longrightarrow} x]$$

(for $b, y \in B$). Let $\mathcal{B}^l$ be the corresponding structure on B. First a simple observation.

12.4 LEMMA. *The canonical assignment $f : A \longrightarrow B$ is a surjective morphism* $\mathcal{A} \longrightarrow \mathcal{B}^l$.

Proof. Suppose that $a \overset{i}{\longrightarrow} x$ for some $a, x \in A$ and label i. Then

$$a \in f(a) \;\;,\;\; x \in f(x) \;\;,\;\; a \overset{i}{\longrightarrow} x$$

and hence $a \overset{i}{\longrightarrow}_l x$, as required. ∎

Let λ be the valuation on $\mathcal{B}^l$ given by

$$b \in \lambda(P) \quad \Leftrightarrow \quad (\exists a \in b)[a \in \alpha(P)]$$

for all variables $P \in \Gamma$ and elements $a \in A$. The values $\lambda(P)$ for other variables P are not important, but for precision we may set

$$\lambda(P) = \emptyset$$

for such P.

12.5 THEOREM. *The assignment*

$$(\mathcal{A}, \alpha) \xrightarrow{\; f \;} (\mathcal{B}^l, \beta)$$

is a Γ-filtration.

Proof. By Lemma 12.4 and the construction of λ the assignment f is a surjective Γ-morphism from $(\mathcal{A}, \alpha)$ to $(\mathcal{B}^l, \lambda)$. It thus remains to verify properties (Var) and (Fil).

By construction we have

$$a \in \alpha(P) \quad \Rightarrow \quad f(a) \in \lambda(P)$$

for all $a \in A$ and variables $P \in \Gamma$. Conversely, if $f(a) \in \lambda(P)$ (for some $a \in A$ and $P \in \Gamma$) then, by definition of λ there is some $x \in f(a)$ with $x \in \alpha(P)$. But then $x \sim a$ so that, by definition of $\sim$, we have $a \in \alpha(P)$, as required to verify (Var).

Finally, to verify (Fil), consider any pair of elements $a, x \in A$ with $f(a) \xrightarrow{\; i \;}_l f(x)$ (in $\mathcal{B}^l$). Then, by definition of $\xrightarrow{\; i \;}_l$, there are $u, v \in A$ with

$$u \sim a \quad , \quad v \sim x \quad , \quad u \xrightarrow{\; i \;} v.$$

Hence, for each formula ϕ with $[i]\phi \in \Gamma$, the definition of $\sim$ gives

$$a \Vdash [i]\phi \Rightarrow u \Vdash [i]\phi \Rightarrow v \Vdash \phi \Rightarrow x \Vdash \phi$$

as required. ∎

12.4   The right-most filtration

For each label i let $\xrightarrow{\; i \;}_r$ be the relation on $B = A/\sim$ given by

$$b \xrightarrow{\; i \;}_r y \Leftrightarrow \left\{ \begin{array}{l} \text{For each formula } \phi \text{ with } [i]\phi \in \Gamma, \\[2mm] (\forall a \in b, x \in y)[a \Vdash [i]\phi \Rightarrow x \Vdash \phi] \end{array} \right\}$$

(for $b, y \in B$). Let $\mathcal{B}^r$ be the corresponding structure on $\mathcal{B}$.

12.6 LEMMA. *The assignment*

$$f : A \longrightarrow B$$

is a surjective morphism $\mathcal{A} \longrightarrow \mathcal{B}^r$.

Proof. Suppose $a \xrightarrow{i} x$ and consider any $u \in f(a), v \in f(x)$. Then

$$u \sim a \quad , \quad x \sim v$$

and for each formula ϕ with $[i]\phi \in \Gamma$ we have

$$u \Vdash [i]\phi \Rightarrow a \Vdash [i]\phi \Rightarrow x \Vdash \phi \Rightarrow v \Vdash \phi$$

so that $f(a) \xrightarrow{i}_r f(x)$ as required. ∎

Let ρ be the valuation on $\mathcal{B}^r$ given by

$$b \in \rho(P) \quad \Leftrightarrow \quad (\forall a \in b)[a \in \alpha(P)]$$

for all variables $P \in \Gamma$ and elements $a \in A$. The values $\lambda(P)$ for other variables P is not important, but for precision we may set

$$\lambda(P) = B$$

for such P.

12.7 THEOREM. *The assignment*

$$(\mathcal{A}, \alpha) \xrightarrow{f} (\mathcal{B}^l, \beta)$$

is a Γ-*filtration.*

Proof. We must verify the conditions (Var) and (Fil).

For (Var) consider any variable $P \in \Gamma$ and element $a \in \alpha(P)$. Then for each $x \in f(a)$, we have $x \sim a$, so that $x \in \alpha(P)$, and hence $f(a) \in \rho(P)$. Conversely, if $f(a) \in \rho(P)$ then, since $a \in f(a)$, we have $a \in \alpha(P)$ as required.

To verify (Fil) consider any $a, x \in A$ with $f(a) \xrightarrow{i}_r f(x)$. Then since $a \in f(a)$ and $x \in f(x)$, for each appropriate formula ϕ we have

$$a \Vdash [i]\phi \quad \Rightarrow \quad x \Vdash \phi$$

as required. ∎

12.5 Filtrations sandwiched

So far we have at least two ways of converting the canonical quotient

$$A \overset{f}{\longrightarrow} B$$

into a Γ-filtration. These are not the only possible constructions but, as we show in this section, all compatible Γ-filtration structures on B are sandwiched between $(\mathcal{B}^l, \lambda)$ and $(\mathcal{B}^r, \rho)$. Before we prove this, a simple observation.

12.8 LEMMA. *The identity map*

$$B \longrightarrow B$$
$$b \longmapsto b$$

provides a Γ-morphism $(\mathcal{B}^l, \lambda) \longrightarrow (\mathcal{B}^r, \rho)$.

Proof. For an arbitrary label i consider any $b, y \in B$ with $b \overset{i}{\longrightarrow}_l y$. By the definition of $\overset{i}{\longrightarrow}_l$ there are $a \in b$ and $x \in y$ with $a \overset{i}{\longrightarrow} x$. Now consider any $u \in b$ and $v \in y$. Then

$$u \sim a \quad , \quad x \sim v$$

so that, for each formula ϕ with $\Box \phi \in \Gamma$ we have

$$u \Vdash [i]\phi \Rightarrow a \Vdash [i]\phi \Rightarrow x \Vdash \phi \Rightarrow v \Vdash \phi$$

so that $b \overset{i}{\longrightarrow}_r y$. This shows that the identity map is a morphism $\mathcal{B}^l \longrightarrow \mathcal{B}^r$.

Next consider any variable $P \in \Gamma$ and element $b \in B$. If $b \in \lambda(P)$ then, by definition of λ, there is some $a \in b$ with $a \in \alpha(P)$, and hence, by the definition of ρ, $b = f(a) \in \rho(P)$. Thus

$$b \in \lambda(P) \Rightarrow b \in \rho(P)$$

which is enough to complete the proof. ∎

Consider now an arbitrary valued structure $(\mathcal{B}, \beta)$ based on the set $B = A/\sim$. We say $(\mathcal{B}, \beta)$ is Γ *sandwiched* if both the following conditions hold.

• For each label i and elements $b, y \in B$ both the implications

$$b \overset{i}{\longrightarrow}_l y \quad \Rightarrow \quad b \overset{i}{\longrightarrow} y \quad \Rightarrow \quad b \overset{i}{\longrightarrow}_r y$$

hold.

• For each variable $P \in \Gamma$ both the inclusions

$$\lambda(P) \subseteq \beta(P) \subseteq \rho(P)$$

hold.

These conditions delimit the range of Γ-filtrations.

12.9 THEOREM. *Let $(\mathcal{B}, \beta)$ be a valued structure based on the quotient set B. Then the canonical assignment $f : A \longrightarrow B$ provides a Γ-filtration if and only if $(\mathcal{B}, \beta)$ is Γ-sandwiched.*

Proof. Suppose first that f is a Γ-filtration.

Consider any label i and elements $b, y \in B$. If $b \xrightarrow{i}_l y$ then there are $a \in b$ and $x \in y$ with $a \xrightarrow{i} x$ hence, since f is a morphism

$$b = f(a) \xrightarrow{i} f(x) = y.$$

Also if $b \xrightarrow{i} y$ then for each $a \in b$ and $x \in y$ we have $f(a) \xrightarrow{i} f(x)$, so that (Fil) ensures that $b \xrightarrow{i}_r y$. Similarly, for each variable $P \in \Gamma$ and elements $a \in A$ and $b \in B$ with $f(a) = b$, we have

$$b \in \lambda(P) \;\Rightarrow\; a \in \alpha(P) \;\Rightarrow\; b \in \beta(P)$$

and

$$b \in \beta(P) \;\Rightarrow\; a \in \alpha(P) \;\Rightarrow\; b \in \rho(P)$$

as required.

This shows that $(\mathcal{B}, \beta)$ is Γ-sandwiched.

Conversely, suppose we know that $(\mathcal{B}, \beta)$ is Γ-sandwiched. Then, for each $a, x \in A$, since f is a morphism $\mathcal{A} \longrightarrow \mathcal{B}^l$, we have

$$a \xrightarrow{i} x \quad \Rightarrow \quad f(a) \xrightarrow{i}_l f(x) \quad \Rightarrow \quad f(a) \xrightarrow{i} f(x).$$

Thus f is a morphism $\mathcal{A} \longrightarrow \mathcal{B}$ and it remains to verify (Var) and (Fil).

The property (Var) follows almost immediately using both ends of the sandwich.

Finally, to verify (Fil), suppose that $f(a) \xrightarrow{i} f(x)$ (for some $a, x \in A$). Then $f(a) \xrightarrow{i}_r f(x)$ and hence, for each appropriate formula ϕ,

$$a \Vdash [i]\phi \quad \Rightarrow \quad x \Vdash \phi$$

as required. ■

12.6   Separated structures

Recall that in Definition 11.4 of Chapter 11 we introduced the semantic equivalence relation $\approx$ between two valued structures. As a particular case of this we can consider the relation on one valued structure $(\mathcal{A}, \alpha)$, i.e. the relation $\approx$ on A given by

$$a \approx x \Leftrightarrow \left\{ \begin{array}{c} \text{For all formulas } \phi, \\ (\mathcal{A}, \alpha, a) \Vdash \phi \Leftrightarrow (\mathcal{A}, \alpha, x) \Vdash \phi \end{array} \right\}$$

for all $a, x \in A$. We say a valued structure is *separated* if this relation is just equality, i.e. if

$$a \approx x \Rightarrow a = x$$

holds for all $a, x \in A$. In this section we observe how filtration constructions produce separated structures.

Thus, fix the usual data of a valued structure $(\mathcal{A}, \alpha)$, a set of formulas Γ with the induced equivalence relation $\sim$ on A, and the quotient set $B = A/\sim$. Let

$$(\mathcal{A}, \alpha) \xrightarrow{f} (\mathcal{B}, \beta)$$

be a filtration where the target structure $(\mathcal{B}, \beta)$ is carried by B. Note that

$$a \sim x \Leftrightarrow f(a) = f(x)$$

holds for all $a, x \in A$.

12.10 LEMMA. *In the circumstances given above, the structure $(\mathcal{B}, \beta)$ is separated.*

Proof. For each $a, x \in A$ we have

$$f(a) \approx f(x) \Rightarrow \left\{ \begin{array}{c} \text{For all formulas } \phi, \\ f(a) \Vdash \phi \Leftrightarrow f(x) \Vdash \phi \end{array} \right\}$$

$$\Rightarrow \left\{ \begin{array}{c} \text{For all formulas } \phi \in \Gamma, \\ f(a) \Vdash \phi \Leftrightarrow f(x) \Vdash \phi \end{array} \right\}$$

$$\Rightarrow \left\{ \begin{array}{c} \text{For all formulas } \phi \in \Gamma, \\ a \Vdash \phi \Leftrightarrow x \Vdash \phi \end{array} \right\}$$

$$\Rightarrow \qquad a \sim x \qquad \Rightarrow \quad f(a) = f(x)$$

where these implications follow by

• the definition of $\approx$,

• restriction,

- the filtration preservation property,

- the definition of $\sim$,

and finally the above remark. ■

12.7   Exercises

12.1 Let Γ be the set of sentences (variable-free formulas). Note that for this set Γ the valuation plays very little part in the notion of a Γ-filtration.
Consider the 13 element monomodal structure $\mathcal{A}$

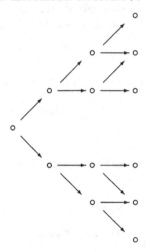

where all the transitions are displayed explicitly (in particular, no node is reflexive).

(a) Determine the equivalence relation $\sim$ on $\mathcal{A}$ induced by Γ.

(b) Show that the left-most and the right-most Γ-filtrations of $\mathcal{A}$ are identical and both have just four elements.

12.2 Let $\mathcal{N}^+$ and $\mathcal{N}^-$ be the two monomodal structures on N (the set of natural numbers) with transition relations given by

$$ a \xrightarrow{\ +\ } b \ \Leftrightarrow\ b = a + 1 \quad , \quad a \xrightarrow{\ -\ } b \ \Leftrightarrow\ a = b + 1. $$

Let Γ be the set of sentences.

(a) Show that the left-most Γ-filtration of $\mathcal{N}^+$ effects a complete collapse to a reflexive point.

(b) Show that the right-most Γ-filtration of $\mathcal{N}^-$ is an isomorphism.

12.3 Consider the infinite monomodal structure $\mathcal{A}$

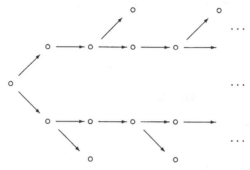

where all the transition are displayed explicitly (in particular, no node is reflexive) Let Γ be the set of sentences.

Construct the left-most Γ-filtration of $\mathcal{A}$.

12.4 Show that the target of a filtration need be neither finite nor separated.

12.5 Let Γ be the set of variables. Let $\mathcal{A}$ be any monomodal structure and let α be any valuation on $\mathcal{A}$ such that for each $a \in A$ there is some $P \in \Gamma$ with $\alpha(P) = \{a\}$. Determine the left-most and right-most Γ-filtrations of $(\mathcal{A}, \alpha)$.

12.6 Suppose the valued structure $(\mathcal{A}, \alpha)$ is both finite and separated.

(a) Show that for each pair of distinct elements a, x of $\mathcal{A}$, there is a formula $\xi_{a,x}$ which distinguishes between a and x in the sense that both

$$(\mathcal{A}, \alpha, a) \Vdash \xi_{a,x} \quad \text{and} \quad (\mathcal{A}, \alpha, x) \Vdash \neg\xi_{a,x}$$

hold.

(b) Show that for each $a \in A$ there is a formula ρ_a such that

$$(\mathcal{A}, \alpha, x) \Vdash \rho_a \iff x = a$$

holds for all $x \in A$.

(c) Show that for each $X \subseteq A$ there is a formula τ_X such that

$$(\mathcal{A}, \alpha, a) \Vdash \tau_X \iff a \in X$$

holds for all $a \in A$.

Let μ be any valuation on $\mathcal{A}$ and let $(\cdot)^\mu$ be the substitution given by

$$P \longmapsto \tau_{\mu(P)}$$

for each variable P.

(d) Show that
$$(\mathcal{A}, \mu, a) \Vdash \phi \Leftrightarrow (\mathcal{A}, \alpha, a) \Vdash \phi^\mu$$
holds for all elements a of $\mathcal{A}$ and formulas ϕ.

12.7 Let Γ be the set of all formulas and let

$$(\mathcal{A}, \alpha) \xrightarrow{\ f\ } (\mathcal{B}, \beta)$$

be a surjective valued morphism.

(a) Show that if f is a zigzag morphism then f is a Γ-filtration.

(b) Show that if $(\mathcal{B}, \beta)$ is finite and separated and f is a Γ-filtration, then f is a zigzag morphism.

Chapter 13

The finite model property

13.1 Introduction

We have seen several examples of completeness results of the form

$$\vdash_S \phi \quad \Leftrightarrow \quad \mathbb{M} \text{ models } \phi$$

where S is a standard system and $\mathbb{M}$ is a suitable class of structures (and ϕ is an arbitrary formula). However, in all of these cases there has been little or no information about the size of the structures in $\mathbb{M}$. In this chapter we investigate the consequences of the existence of such an equivalence for a class $\mathbb{M}$ of finite structures.

13.2 The fmp explained

Let S be any standard system and consider the following three classes of finite structures.

$\mathbb{F}$ = the class of finite valued structures which model S
$\mathbb{G}$ = the class of finite (unadorned) structures which model S
$\mathbb{H}$ = the class of separated valued structures which model S

For any of these classes $\mathbb{K}$, let $Th(\mathbb{K})$ be the set of all formulas ϕ modelled by $\mathbb{K}$. Clearly

$$Th(\mathbb{F}) \subseteq Th(\mathbb{H}) \quad , \quad Th(\mathbb{F}) \subseteq Th(\mathbb{G})$$

(since $\mathbb{H} \subseteq \mathbb{F}$ and each valuation on a member of $\mathbb{G}$ produces a member of $\mathbb{F}$).

Much of the usefulness of the fmp is due to the following result.

13.1 THEOREM. *The three sets*

$$Th(\mathbb{F}) \quad , \quad Th(\mathbb{G}) \quad , \quad Th(\mathbb{H})$$

are equal.

169

Proof. We show that

$$Th(\mathbb{G}) \subseteq Th(\mathbb{H}) \subseteq Th(\mathbb{F})$$

which, with the above observation, gives the required result.

For the first inclusion consider any $(\mathcal{A}, \alpha) \in \mathbb{H}$. From Exercise 12.1 we know that for each valuation μ on $\mathcal{A}$ and formula ϕ there is an appropriate substitution instance ϕ^μ of ϕ such that

$$(\mathcal{A}, \mu) \Vdash^v \phi \quad \Leftrightarrow \quad (\mathcal{A}, \alpha) \Vdash^v \phi^\mu.$$

In particular, since the axioms of S are closed under substitution and $(\mathcal{A}, \alpha)$ models S, we see that $(\mathcal{A}, \mu)$ also models S. Thus $\mathcal{A}$ models S and hence $\mathcal{A} \in \mathbb{G}$.

This shows that for each formula ϕ,

$$\phi \in Th(\mathbb{G}) \quad \Rightarrow \quad \mathcal{A} \Vdash^u \phi \quad \Rightarrow \quad (\mathcal{A}, \alpha) \Vdash^v \phi$$

and hence (since $(\mathcal{A}, \alpha)$ is an arbitrary member of $\mathbb{H}$) we have

$$\phi \in Th(\mathbb{G}) \quad \Rightarrow \quad \phi \in Th(\mathbb{H})$$

as required.

For the second inclusion consider any member $(\mathcal{A}, \alpha)$ of $\mathbb{F}$ and let $\approx$ be the semantic equivalence relation on A given by

$$a \approx b \Leftrightarrow \left\{ \begin{array}{c} \text{For all formulas } \phi, \\ a \Vdash \phi \quad \Leftrightarrow \quad b \Vdash \phi \end{array} \right\}$$

(for $a, b \in A$). We know we may slice $(\mathcal{A}, \alpha)$ by $\approx$ to produce a filtration

$$(\mathcal{A}, \alpha) \xrightarrow{\ f\ } (\mathcal{B}, \beta)$$

for which

$$(\mathcal{A}, \alpha, a) \Vdash \phi \quad \Leftrightarrow \quad (\mathcal{B}, \beta, f(a)) \Vdash \phi$$

for all formulas ϕ and $a \in A$. We know that $\mathcal{B}$ is separated, and hence $(\mathcal{B}, \beta) \in \mathbb{H}$. This is enough to give the required inclusion. ∎

Now let

$$\mathsf{S}(fin) = Th(\mathbb{F}) = Th(\mathbb{G}) = Th(\mathbb{H})$$

so that

$$\vdash_\mathsf{S} \phi \quad \Rightarrow \quad \phi \in \mathsf{S}(fin)$$

(for all formulas ϕ).

13.2 DEFINITION. We say the system S has the *f(inite) m(odel) p(roperty)* if the above implication is, in fact, an equivalence. ■

Although it is not within the scope of this book, it is worth saying a few words about the consequences of the fmp concerning decidability.

Most formal system S that we are concerned with are such that the set S(*all*) of all formula ϕ for which

$$\vdash_S \phi$$

is automatically recursive enumerable. This is the case when S is finitely axiomatizable. If such a system also has the fmp then, in fact, it is decidable, i.e. the set S(*all*) is recursive. To see this we use the equality S(*all*) = S(*fin*) and the fact that the complement of S(*fin*) is recursively enumerable. To enumerate the complement of S(*fin*) first begin to enumerate all finite structures. Using the fact that S is finitely axiomatizable we can extract from this list an enumeration of all finite models of S. We then enumerate all the formulas which are not modelled by at least one structure in this second list. This list of formulas is the complement of S(*fin*).

We formally record this fact.

13.3 THEOREM. *Let S be a finitely axiomatizable system with the fmp. Then S is decidable.*

Let us now continue with our study of the fmp. First we observe that it is not necessary to use all finite models of a system.

13.4 LEMMA. *Suppose S is a standard system and* **M** *is any class of finite structures such that the equivalence*

$$\vdash_S \phi \quad \Leftrightarrow \quad \mathbb{M} \ models \ \phi$$

holds. Then S has the fmp.

Proof. All the structures in **M** model S so that either

$$\mathbb{M} \subseteq \mathbb{F} \quad or \quad \mathbb{M} \subseteq \mathbb{G}$$

depending on whether the structures are valued or not. But then either

$$S(fin) = Th(\mathbb{F}) \subseteq Th(\mathbb{M}) \quad or \quad S(fin) = Th(\mathbb{G}) \subseteq Th(\mathbb{M})$$

so that for each formula ϕ

$$\phi \in S(fin) \quad \Rightarrow \quad \phi \in Th(\mathbb{M}) \quad \Rightarrow \quad \vdash_S \phi$$

which gives the required result. ■

The fmp is another way of obtaining completeness.

13.5 THEOREM. *Let* S *be a standard system with the fmp. Then* S *is Kripke-complete.*

Proof. Let $\mathbb{M}$ be the class of all the unadorned models of S. Then $\mathbb{G} \subseteq \mathbb{M}$ and for each formula ϕ

$$\mathbb{M} \text{ models } \phi \quad \Rightarrow \quad \mathbb{G} \text{ models } \phi \quad \Rightarrow \quad \vdash_S \phi$$

where the second implication holds by the fmp. This gives the required Kripke-completeness. ∎

Theorems 13.3 and 13.5 show that finitely axiomatizable systems with the fmp are rather pleasant. Thus it is worth looking at some examples of such systems.

13.3   The classic systems have the fmp

The classic systems are those monomodal systems whose axioms are the various combinations of the shapes

$$\text{D, T, B, 4, 5.}$$

(There are 15 such systems.) We use several different filtrations to show that each of these systems has the fmp.

To illustrate the method used let us begin with the minimal system K.

13.6 THEOREM. *The system* K *has the fmp.*

Proof. Consider any formula ϕ such that

$$\text{not}[\vdash_S \phi].$$

It suffices to produce a finite valued structure which models K but which does not model this formula ϕ.

We know there is some valued structure $(\mathcal{A}, \alpha)$ which models K but does not model ϕ. This model may not be finite so we produce the required model by taking a slice of $(\mathcal{A}, \alpha)$.

Let Γ be the set of subformulas of ϕ and let

$$(\mathcal{A}, \alpha) \xrightarrow{\ f\ } (\mathcal{B}, \beta)$$

be any Γ-filtration (say the left-most one). Since Γ is finite, the structure $(\mathcal{B}, \beta)$ is finite and (trivially) models K. Also, for each $\gamma \in \Gamma$ and $a \in A$,

$$(\mathcal{A}, \alpha, a) \Vdash \gamma \quad \Leftrightarrow \quad (\mathcal{B}, \beta, f(a)) \Vdash \gamma.$$

Finally, there is some $a \in A$ with $a \Vdash \neg\phi$ and hence (since $\phi \in \Gamma$) we have $f(a) \Vdash \neg\phi$ so that $(\mathcal{B}, \beta)$ is the required model of K which does not model ϕ. ■

We now have to deal with the various axioms D, T, B, 4, and 5. For this we use a mixture of correspondence properties and preservation properties. First a rather trivial result.

13.7 LEMMA. *Suppose*

$$\mathcal{A} \xrightarrow{f} \mathcal{B}$$

is a surjective morphism. Then both

(d) if $\mathcal{A}$ is serial then so is $\mathcal{B}$

(t) if $\mathcal{A}$ is reflexive then so is $\mathcal{B}$

hold.

Proof. (d) Consider any $b_1 \in B$. We must produce some $b_2 \in B$ with $b_1 \longrightarrow b_2$. Since f is surjective there is at least one $a_1 \in A$ with $f(a_1) = b_1$. Since $\mathcal{A}$ is serial there is some $a_2 \in A$ with $a_1 \to a_2$. Since f is a morphism this gives $f(a_1) \longrightarrow f(a_2)$ so we may set $b_2 = f(a_2)$.

(t) This is proved in a similar way. ■

This preservation result is sufficient to show that KD and KT have the fmp. However it is not strong enough to deal with the other axioms because other properties are not preserved by mere surjective morphisms. We need a tighter form of morphism, and this is where filtrations are useful. As an illustration let us see how we handle the shape B.

13.8 LEMMA. *Suppose*

$$(\mathcal{A}, \alpha) \xrightarrow{f} (\mathcal{B}, \beta)$$

is a left-most filtration where $\mathcal{A}$ is symmetric. Then $\mathcal{B}$ is also symmetric.

Proof. Consider any $b_1, b_2 \in B$ with $b_1 \longrightarrow b_2$. Since the filtration is left-most there are $a_1, a_2 \in A$ with

$$f(a_1) = b_1 \quad , \quad f(a_2) = b_2 \quad , \quad a_1 \longrightarrow a_2.$$

Since $\mathcal{A}$ is symmetric, we have $a_2 \longrightarrow a_1$ and hence, again since each filtration is a morphism, we have $b_2 \longrightarrow b_1$, as required. ■

We can now extend our class of examples of systems with the fmp.

13.9 THEOREM. *Let* S *be any of the 6 systems whose axioms are taken from*

$$D, \ T, \ B.$$

Then S *has the fmp.*

Proof. We know that there is a structural property $\parallel S \parallel$ such that for each structure $\mathcal{A}$

$$\mathcal{A} \text{ models S} \quad \Leftrightarrow \quad \mathcal{A} \text{ has } \parallel S \parallel .$$

Also, from Lemmas 13.7 and 13.8, we know that for each left-most filtration

$$(\mathcal{A}, \alpha) \ \overset{f}{\longrightarrow} \ (\mathcal{B}, \beta)$$

if $\mathcal{A}$ has $\parallel S \parallel$ then $\mathcal{B}$ has $\parallel S \parallel$.

Now consider any formula ϕ with

$$\text{not}[\vdash_S \phi].$$

Since S is Kripke-complete, there is some structure $\mathcal{A}$ with property $\parallel S \parallel$ and which does not model ϕ, i.e. there is a valuation α on $\mathcal{A}$ such that $(\mathcal{A}, \alpha)$ does not model ϕ.

Let Γ be the set of subformulas of ϕ and let f, as above, be the left-most Γ-filtration of $(\mathcal{A}, \alpha)$. Note that $\mathcal{B}$ is is finite (since Γ is) and $\mathcal{B}$ has property $\parallel S \parallel$, and hence models S. Finally, since $\phi \in \Gamma$, we know, by the filtration property, that $(\mathcal{B}, \beta)$ does not model ϕ, from which we obtain the required result. ∎

We now turn to the shape 4 which, of course, characterizes transitivity. To deal with this we need a custom built filtration.

As usual, let Γ be any set of formulas which is closed under subformulas. Also let $(\mathcal{A}, \alpha)$ be any structure, and consider the canonical Γ-quotient

$$A \ \overset{f}{\longrightarrow} \ B$$

where

$$f(a_1) = f(a_2) \quad \Leftrightarrow \quad (\forall \gamma \in \Gamma)[a_1 \Vdash \gamma \Leftrightarrow a_2 \Vdash \gamma]$$

(for $a_1, a_2 \in A$). We convert B into a structure using the transition relation $\longrightarrow$ given by

$$b_1 \longrightarrow b_2 \quad \Leftrightarrow \quad \left\{ \begin{array}{l} \text{For all } a_1 \in b_1, a_2 \in b_2 \text{ and} \\ \text{formulas } \phi \text{ with } \Box\phi \in \Gamma, \\ a_1 \Vdash \Box\phi \ \Rightarrow \ a_2 \Vdash \phi \wedge \Box\phi. \end{array} \right.$$

We impose on $\mathcal{B}$ the usual valuation β given by

$$b \in \beta(P) \quad \Leftrightarrow \quad (\exists a \in b)[a \in \alpha(P)]$$

(for $P \in \Gamma$ with other values of β unimportant).

This construction is sometimes known as the Lemmon filtration.

13.10 LEMMA. *Let $(\mathcal{A}, \alpha)$ be any valued structure with $\mathcal{A}$ transitive. Then the above construction produces a Γ-filtration*

$$(\mathcal{A}, \alpha) \xrightarrow{f} (\mathcal{B}, \beta)$$

for which $\mathcal{B}$ is also transitive.

Proof. We verify the various conditions in turn.

To show that f is a morphism consider any $a_1, a_2 \in A$ with $a_1 \to a_2$. We must show that $f(a_1) \to f(a_2)$. To this end consider any $x_1 \in f(a_1)$ and $x_2 \in f(a_2)$. Then for each formula ϕ with $\Box\phi \in \Gamma$ we have

$$
\begin{aligned}
x_1 \Vdash \Box\phi &\Rightarrow a_1 \Vdash \Box\phi && \text{(since } f(x_1) = f(a_1)\text{)} \\
&\Rightarrow a_1 \Vdash \Box\phi \wedge \Box^2\phi && \text{(since } \mathcal{A} \text{ is transitive)} \\
&\Rightarrow a_2 \Vdash \phi \wedge \Box\phi && \text{(since } a_1 \to a_2\text{)} \\
&\Rightarrow x_2 \Vdash \phi \wedge \Box\phi && \text{(since } f(a_2) = f(x_2)\text{)}
\end{aligned}
$$

as required.

The three conditions (Sur, Var, Fil) hold by construction.

Finally we must show that $\mathcal{B}$ is transitive. Thus, consider any $b_1, b_2, b_3 \in B$ with

$$b_1 \longrightarrow b_2 \longrightarrow b_3$$

and let $a_i \in b_i$ for $i = 1, 2, 3$. Then for each formula ϕ with $\Box\phi \in \Gamma$ we have

$$
\begin{aligned}
a_1 \Vdash \Box\phi &\Rightarrow a_2 \Vdash \phi \wedge \Box\phi \\
&\Rightarrow a_2 \Vdash \Box\phi \quad \Rightarrow a_3 \Vdash \phi \wedge \Box\phi
\end{aligned}
$$

so that $b_1 \longrightarrow b_3$ as required. ∎

This result with Lemma 13.7 is enough to obtain the following result.

13.11 THEOREM. *The three systems*

$$\mathsf{K4} \quad , \quad \mathsf{KD4} \quad , \quad \mathsf{KT4} = \mathsf{S4}$$

all have the fmp.

Next we look at the combination of transitivity and symmetry, i.e. the models of **KB4**. Again we need a custom built filtration. Thus consider the usual data of a finite set of formulas Γ which is closed under subformulas, let $(\mathcal{A}, \alpha)$ be any valued structure, and consider the canonical Γ-quotient

$$A \xrightarrow{f} B$$

We convert B into a structure $\mathcal{B}$ using a transition relation $\longrightarrow$ and then impose the usual valuation β on $\mathcal{B}$. For transitivity and symmetry the appropriate relation is defined as follows.

$b_1 \longrightarrow b_2 \quad \Leftrightarrow \quad$ For all $a_1 \in b_1$ and $a_2 \in b_2$ and formulas ϕ with $\Box \phi \in \Gamma$,
$$a_1 \Vdash \Box \phi \quad \Rightarrow \quad a_2 \Vdash \phi \wedge \Box \phi$$
and
$$a_2 \Vdash \Box \phi \quad \Rightarrow \quad a_1 \Vdash \phi \wedge \Box \phi$$

I will leave the proof of the following to you.

13.12 LEMMA. *Let* $(\mathcal{A}, \alpha)$ *be any valued structure with* $\mathcal{A}$ *transitive and symmetric. Then the above construction produces a* Γ-*filtration*

$$(\mathcal{A}, \alpha) \xrightarrow{\;f\;} (\mathcal{B}, \beta)$$

for which $\mathcal{B}$ *is also transitive and symmetric.*

These results are enough to prove the following result.

13.13 THEOREM. *Let* S *be any of the 11 standard formal system whose axioms are made up of various combinations of the shapes* D, T, B, *and* 4. *Then* S *has the fmp.*

We now consider the systems formed by extending the above 11 systems by the addition of the axiom 5. This gives us four new systems K5, KD5 and K45, kD45.

The axiom 5 alone, which captures the euclidean property, is a little more delicate to deal with. We need a bit of a preamble.

Starting from any set Γ of formulas, which, as usual, we assume is finite and closed under subformulas, we first set

$$\Box \Gamma = \{\, \Box \phi \mid \phi \in \Gamma \,\} \quad , \quad \Diamond \Gamma = \{\, \Diamond \phi \mid \phi \in \Gamma \,\}$$

and then set
$$\Gamma^{\bullet} = \Gamma \cup \Box \Gamma \cup \Diamond \Gamma.$$

Thus $\Gamma \subseteq \Gamma^{\bullet}$ and $\Gamma^{\bullet}$ is also finite and close under subformulas. We now iterate this construction to get

$$\Gamma_0 = \Gamma \quad , \quad \Gamma_{r+1} = (\Gamma_r)^{\bullet}$$

for each $r < \omega$, and then set

$$\Gamma^* = \bigcup \{\Gamma_r \mid r < \omega\}.$$

In general Γ^* is closed under subformulas but need not be finite. However, when we work relative to the base system K5, we can retrieve finiteness.

By Exercise 8.6 of Chapter 8 we know that modulo K5 each formula has just finitely many modal variants. More precisely, we know that for each $\psi \in \Gamma^*$ there is some $\theta \in \Gamma^{\bullet\bullet}$ with

$$\vdash_{K5} \psi \leftrightarrow \theta$$

so that Γ^* contains no more information than $\Gamma^{\bullet\bullet}$.
Consider any valued structure $(\mathcal{A}, \alpha)$ and let

$$A \xrightarrow{f} B$$

be the canonical $\Gamma^{\bullet\bullet}$-quotient. Let $\longrightarrow$ be the transition relation on B given
by

$$b_1 \longrightarrow b_2 \Leftrightarrow \begin{cases} \text{For each } a_1 \in b_1, a_2 \in b_2 \text{ and formula } \phi \in \Gamma^{\bullet\bullet}, \\ a_1 \Vdash \Box\phi \Rightarrow a_2 \Vdash \phi \\ \text{and} \\ a_1 \Vdash \Diamond\phi \Leftarrow a_2 \Vdash \phi. \end{cases}$$

Note the direction of the second implication in this definition.
This gives us a structure $\mathcal{B}$ on which we impose the usual valuation.

13.14 THEOREM. *When the structure $\mathcal{A}$ is euclidean, the assignment f produces a $\Gamma^{\bullet\bullet}$-filtration*

$$(\mathcal{A}, \alpha) \xrightarrow{f} (\mathcal{B}, \beta)$$

where $\mathcal{B}$ is also euclidean.

Proof. We verify first that f is an unadorned morphism. Thus consider
any $x_1, x_2 \in A$ with

$$x_1 \longrightarrow x_2$$

and any $a_1 \in f(x_1), a_2 \in f(x_2)$. For each $\phi \in \Gamma^{\bullet\bullet}$ there are $\psi, \theta \in \Gamma^{\bullet\bullet}$ with

$$\vdash_{\mathsf{K5}} \Box\phi \leftrightarrow \psi \quad , \quad \vdash_{\mathsf{K5}} \Diamond\phi \leftrightarrow \theta$$

and (since $\mathcal{A}$ models K5) both of these equivalences hold in $\mathcal{A}$. Hence

$$\begin{aligned} a_1 \Vdash \Box\phi &\Rightarrow a_1 \Vdash \psi \\ &\Rightarrow x_1 \Vdash \psi \\ &\Rightarrow x_1 \Vdash \Box\phi \\ &\Rightarrow x_2 \Vdash \phi \quad \Rightarrow a_2 \Vdash \phi \end{aligned}$$

where these implications follow by

- the above equivalence,

- the $\Gamma^{\bullet\bullet}$-induced equivalence,

- the above equivalence,

- the transition $x_1 \longrightarrow x_2$,

- the $\Gamma^{\bullet\bullet}$-induced equivalence.

A similar argument shows that

$$a_1 \Vdash \Diamond \phi \;\Leftarrow\; a_2 \Vdash \phi$$

and hence we get

$$f(x_1) \longrightarrow f(x_2)$$

to verify the morphism property.

The remaining properties required for a filtration are straight forward. Finally, we must show that the target structure $\mathcal{B}$ is euclidean. To this end consider any elements b_1, b_2, b_3 of $\mathcal{B}$ with

We show that $b_2 \longrightarrow b_3$. Thus, consider any

$$a_1 \in b_1 \quad, \quad a_2 \in b_2 \quad, \quad a_3 \in b_3$$

and any formula $\phi \in \Gamma^{\bullet\bullet}$. Then

$$a_2 \Vdash \Box \phi \;\Rightarrow\; a_1 \Vdash \Diamond \Box \phi \;\Rightarrow\; a_1 \Vdash \Box \phi \;\Rightarrow\; a_3 \Vdash \phi$$

where these implications follow by

- the **K5**-equivalence trick used above and the definition of $b_1 \longrightarrow b_2$,

- the variant $\Diamond \Box \phi \to \Box \phi$ of the 5 axiom,

- the definition of $b_1 \longrightarrow b_3$.

A similar argument shows that

$$a_2 \Vdash \Diamond \phi \;\Leftarrow\; a_3 \Vdash \phi$$

and hence $b_2 \longrightarrow b_3$, as required. ■

A routine argument now gives the following.

13.15 THEOREM. *The two systems* **K5** *and* **KD5** *have the fmp.*

Finally, it remains to deal with **K45** and **KD45** which are the systems that capture transitive and euclidean (and serial) structures. Here we use the relation

$b_1 \longrightarrow b_2$ $\Leftrightarrow$ For all $a_1 \in b_1$ and $a_2 \in a_2$ and formulas ϕ with $\Box \phi \in \Gamma$,
$$a_1 \Vdash \Box \phi \quad \Rightarrow \quad a_2 \Vdash \phi \wedge \Box \phi$$
and
$$a_1 \Vdash \Box \phi \quad \Leftrightarrow \quad a_2 \Vdash \Box \phi$$

and then impose the usual valuation on the structure.

13.16 LEMMA. *Let $(\mathcal{A}, \alpha)$ be any valued structure with $\mathcal{A}$ transitive and euclidean. Then the above construction produces a Γ-filtration*

$$(\mathcal{A}, \alpha) \xrightarrow{\ f\ } (\mathcal{B}, \beta)$$

for which $\mathcal{B}$ is also transitive and euclidean.

Proof. Assuming we have already shown that the construction does give a Γ-filtration, let us see why it preserves the euclidean property.

Thus consider any $b_1, b_2, b_3 \in B$ with

We must show that $b_2 \longrightarrow b_3$. To this end consider any $a_1 \in b_1, a_2 \in b_2, a_3 \in b_3$ and consider any formula ϕ with $\Box \phi \in \Gamma$. Then the definition of $\longrightarrow$ on B gives

$$a_2 \Vdash \Box \phi \quad \Rightarrow \quad a_1 \Vdash \Box \phi \quad \Rightarrow \quad a_2 \Vdash \phi$$

and

$$a_2 \Vdash \Box \phi \Leftrightarrow a_1 \Vdash \Box \phi \Leftrightarrow a_3 \Vdash \Box \phi$$

so that $b_2 \to b_3$, as required. ■

Putting all of this together with various earlier results we have all the ingredients for the proof of the following.

13.17 THEOREM. *Let S be any standard system whose axioms are made up of various combinations of the shapes D, T, B, 4, and 5. Then S*

- *is finitely axiomatizable,*

- *is Kripke-complete,*

- *has the fmp,*

- *is decidable,*

and hence S is very pleasant.

I will leave you to provide the various details.

13.4   The basic temporal system has the fmp

So far we have used the filtration technique to obtain the fmp only for mono-modal systems. However, the technique can also be used with polymodal systems, but some of the details are a little more intricate. As an example of this let us show that the basic temporal system TEMP has the fmp.

Recall that, from Chapter 8, Section 8.4 the models of this system are the temporal structures, i.e. the structures

$$\mathcal{A} = (A, \xrightarrow{+}, \xrightarrow{-})$$

where the two relations are converses of each other and both are transitive. Fix such a structure together with a valuation α on $\mathcal{A}$. Fix also some set Γ of formulas which is closed under subformulas. (In any application this will be the set of subformulas of some given formula.) The proof of the fmp boils down to the construction of a Γ-filtration

$$(\mathcal{A}, \alpha) \xrightarrow{f} (\mathcal{B}, \beta)$$

where $\mathcal{B}$ is also a temporal structure.

To do this consider the canonical Γ-quotient

$$A \xrightarrow{f} B.$$

We will furnish B with two transition relations $\xrightarrow{+}, \xrightarrow{-}$ converting B into the required temporal structure $\mathcal{B}$, and then impose the usual valuation β on $\mathcal{B}$. Thus define $\xrightarrow{+}, \xrightarrow{-}$ on B as follows.

$b_1 \xrightarrow{+} b_2 \quad \Leftrightarrow \quad$ For all $a_1 \in b_1, a_2 \in b_2$ and for all formulas ϕ,
$\quad\quad$ (+) if $[+]\phi \in \Gamma$ then
$\quad\quad\quad\quad a_1 \Vdash [+]\phi \quad \Rightarrow \quad a_2 \Vdash \phi \wedge [+]\phi$
$\quad\quad$ (−) if $[-]\phi \in \Gamma$ then
$\quad\quad\quad\quad a_2 \Vdash [-]\phi \quad \Rightarrow \quad a_1 \Vdash \phi \wedge [-]\phi$

$b_1 \xrightarrow{-} b_2 \quad \Leftrightarrow \quad$ For all $a_1 \in b_1, a_2 \in b_2$ and for all formulas ϕ,
$\quad\quad$ (−) if $[-]\phi \in \Gamma$ then
$\quad\quad\quad\quad a_1 \Vdash [-]\phi \quad \Rightarrow \quad a_2 \Vdash \phi \wedge [-]\phi$
$\quad\quad$ (+) if $[+]\phi \in \Gamma$ then
$\quad\quad\quad\quad a_2 \Vdash [+]\phi \quad \Rightarrow \quad a_1 \Vdash \phi \wedge [+]\phi$

Let

$$\mathcal{B} = (B, \xrightarrow{+}, \xrightarrow{-}).$$

We need to check several things.

13.18 LEMMA. *The function f is a morphism $\mathcal{A} \longrightarrow \mathcal{B}$.*

Proof. Consider first any $a_1, a_2 \in A$ with $a_1 \xrightarrow{+} a_2$. We must show that $f(a_1) \xrightarrow{+} f(a_2)$. To this end consider any $x_1 \in f(a_1), x_2 \in f(a_2)$ and any formula ϕ. Then:

    if $[+]\phi \in \Gamma$ then

$$
\begin{array}{lll}
x_1 \Vdash [+]\phi & \Rightarrow a_1 \Vdash [+]\phi & (\text{since } x_1 \simeq a_1) \\
& \Rightarrow a_1 \Vdash [+]\phi \wedge [+]^2\phi & (\text{by axiom 4}) \\
& \Rightarrow a_2 \Vdash \phi \wedge [+]\phi & (\text{since } a_1 \xrightarrow{+} a_2) \\
& \Rightarrow x_2 \Vdash \phi \wedge [+]\phi & (\text{since } a_2 \simeq x_2)
\end{array}
$$

and

    if $[-]\phi \in \Gamma$ then

$$
\begin{array}{lll}
x_2 \Vdash [-]\phi & \Rightarrow a_2 \Vdash [-]\phi & (\text{since } x_2 \simeq a_2) \\
& \Rightarrow a_2 \Vdash [-]\phi \wedge [-]^2\phi & (\text{by axiom 4}) \\
& \Rightarrow a_1 \Vdash \phi \wedge [-]\phi & (\text{since } a_2 \xrightarrow{+} a_1) \\
& \Rightarrow x_1 \Vdash \phi \wedge [-]\phi & (\text{since } a_1 \simeq x_1)
\end{array}
$$

which gives the first required result.

A similar argument also shows that

$$
a_1 \xrightarrow{-} a_2 \quad \Rightarrow \quad f(a_1) \xrightarrow{-} f(a_2)
$$

to complete the proof. ∎

    This with a couple of trivial observations shows that we have constructed a Γ-filtration

$$
(\mathcal{A}, \alpha) \xrightarrow{\ f\ } (\mathcal{B}, \beta)
$$

and so it remains to show that $\mathcal{B}$ is a temporal structure.

Clearly (by definition) the two relations $\xrightarrow{+}, \xrightarrow{-}$ on B are converses. Thus the final required piece of information is provided by the following.

13.19 LEMMA. *Both the relations of $\mathcal{B}$ are transitive.*

Proof. Consider any $b_1, b_2, b_3 \in B$ with

$$
b_1 \xrightarrow{+} b_2 \xrightarrow{+} b_3
$$

and consider any $a_1 \in b_1, a_2 \in b_2, a_3 \in b_3$. Then, for each formula ϕ,

    (+) if $[+]\phi \in \Gamma$ then

$$
\begin{array}{lll}
a_1 \Vdash [+]\phi & \Rightarrow a_2 \Vdash \phi \wedge [+]\phi & \\
& \Rightarrow a_2 \Vdash [+]\phi & \Rightarrow a_3 \Vdash \phi \wedge [+]\phi
\end{array}
$$

and

$(-)$ if $[-]\phi \in \Gamma$ then
$$a_1 \Vdash [-]\phi \;\Rightarrow\; a_2 \Vdash \phi \wedge [-]\phi$$
$$\Rightarrow\; a_2 \Vdash [-]\phi \quad\Rightarrow\; a_3 \Vdash \phi \wedge [-]\phi$$

which shows that $b_1 \xrightarrow{+} b_2$ and hence $\xrightarrow{+}$ is transitive.
A similar argument shows that $\xrightarrow{-}$ is also transitive. ∎

I now expect you to fill in the details of the following result.

13.20 THEOREM. *The basic temporal system*

- *is finitely axiomatizable,*

- *is Kripke-complete,*

- *has the fmp,*

- *is decidable,*

and hence is very pleasant.

13.5   Exercises

13.1 Let S be a standard formal system axiomatized by a single sentence. By Exercise 10.1 of Chapter 10 we know that S is canonical. Now show that S has the fmp.

13.2 Show that a left-most filtration of a pathetic structure has a pathetic target. Hence show that

$$KP, KDP, KTP, KBP$$

all have the fmp.

13.3 Recall that the shape

$$[k]\phi \to [i]\,[j]\phi$$

(for arbitrary ϕ) captures a certain composition property. Let $\mathcal{A}$ be the 4-element structure

$$\circ \xrightarrow{i} \circ \qquad \circ \xrightarrow{j} \circ$$

consisting of one i-transition, one j-transition, and no k-transitions. By considering a certain valuation on $\mathcal{A}$, show that the above formula need not be preserved by left-most filtrations.

13.4 Filtrations are useful because of their preservation properties.

(a) Show that for arbitrary labels i and j the shapes

(i) $[i]\phi \to \langle j \rangle \phi$ (ii) $[i]\phi \to [j]\phi$

(iii) $\phi \to [j]\langle k \rangle \phi$ (iv) $[i]\phi \to [j]\langle k \rangle \phi$

are preserved by left-most filtrations.

(b) Consider that particular case of (ii) where i labels $\square$ and j labels $\square^2$. Thus the shape $\square\phi \to \square^2\phi$, i.e. transitivity, is preserved by left-most filtrations. What, if anything, is wrong with this argument?

13.5   Consider the formal system KE of Exercise 10.2 of Chapter 10.

(a) Show that KE has the fmp.

(b) Show that the extensions KBE and KDE have the fmp.

(c) Using a modified Lemmon filtration, show that KE4 has the fmp.

13.6   Consider the formal system KF of Exercise 10.2 of Chapter 10.

(a) Show that KF, KBF and KDF have the fmp.

(b) Show that KF4 has the fmp.

13.7   Consider the two shapes E and F of Exercise 10.2 of Chapter 10.

(a) Show that KE5 has the fmp.

(b) What can you say about KF5?

13.8   Recall the notion of a *good* structure introduced in Exercise 4.12, namely, a
monomodal structure which is transitive and serial and such that each of its elements can see a reflexive element. Suppose the monomodal valued structure $(\mathcal{A}, \alpha)$ is transitive and serial (but not necessarily good). For an arbitrary formula ϕ let Γ be the set of subformulas of ϕ and let

$$(\mathcal{A}, \alpha) \longrightarrow (\mathcal{B}, \beta)$$

be the Lemmon Γ-filtration.

(a) Show that $\mathcal{B}$ is good.

(b) Show there is no formula whose models are precisely the class of good structures.

Part V

More advanced material

We have isolated three important properties of standard formal systems: being canonical, and having the fmp, both of which imply the third property of being Kripke-complete. We have also seen may examples of systems with these properties. In this part we look more closely at the connection between these properties.

In each of Chapters 14 and 15 we look at a system which has the fmp but is not canonical. Both of these systems have independent interest. In Chapter 16 we consider a system which is canonical but does not have the fmp. This system is custom built to have these properties but may, in time, be found to have interest in its own right. Finally, in Chapter 17, we look at two systems, one of which has all three properties and the other having none of the properties. Furthermore, these two systems have precisely the same class of unadorned models.

Taken as a whole these four chapters hint at some of the complexities that can arise in modal logic.

Chapter 14

SLL logic

14.1 Introduction

Dynamic logic is an extension of polymodal logic designed to aid the analysis of the content and shape of programming languages. The labels of the signature are thought of as particular programs. Each transition structure is thought of as a machine on which these programs may be executed, with the elements of the structure corresponding to the various possible states (or configurations) of the machine. Given two such states a and b and a program (label) i,

$$a \xrightarrow{i} b$$

indicates that a run of the program i in the machine when it is configuration a may result in a transition to configuration b. (The 'may' here allows for non-deterministic runs, so there may be other resulting configurations as well. Executions of the program i are deterministic precisely when the relation $\xrightarrow{i}$ is (partially) functional.) Formulas of the corresponding modal language express properties of the machine states. So a state will satisfy $[i]\phi$ if after any execution of program i from the state, the resulting state satisfies ϕ.

So far this is just a matter of using different words for familiar notions, however dynamic logic also has some additional facilities.

Firstly, programs can be combined in various ways, and this imposes an algebraic structure on the set of labels which must be reflected in the logic. For instance, given two programs i and j we may combine them sequentially or non-deterministically

$$ij \quad , \quad i \cup j.$$

Here ij is the program which, when run, first executes i and then executes j; and $i \cup j$ is the program which, when run, will execute either i or j non-deterministically. Thus we see that both

$$[k]\phi \leftrightarrow [i][j]\phi \quad \text{and} \quad [l]\phi \leftrightarrow [i]\phi \wedge [j]\phi$$

where $k = ij$ and $l = i \cup j$, ought to be part of the supporting logic.

A less trivial example to handle is caused by an undetermined repetition of a program. For a program i let i^* be the program which executes i an indefinite, but finite, number of times. Thus, morally, i^* is the non-deterministic combination

$$\emptyset \cup i \cup ii \cup iii \cup \cdots$$

(where $\emptyset$ is the program which does nothing), and we would like

$$\text{``}[k]\phi \ \leftrightarrow \ \phi \wedge [i]\phi \wedge [ii]\phi \wedge [iii]\phi \wedge \cdots \text{''}$$

where $k = i^*$, to be part of the supporting logic. Unfortunately the right hand side of this is an infinite conjunction, so we can not regard it as part of the modal language. In this chapter we will see how the desired effect can be achieved in a different way.

As I have remarked already, each formula θ expresses a property which a state may or may not have. But this means it is possible to have a program

$$\theta?$$

which, when executed at a state, will test whether or not that state satisfies θ. More precisely, we want

$$a \xrightarrow{\ \theta?\ } b$$

to hold when

$$a \text{ satisfies } \theta \quad \text{and} \quad a = b$$

so that

$$[\theta?]\phi \ \leftrightarrow \ (\theta \to \phi)$$

should be part of the supporting logic. This kind of axiom is not hard to handle but the whole construction does introduce a new order of complexity.

Formulas are constructed from labels (and other syntactic entities). Usually these labels are atomic objects, but now we have a method of constructing new labels from formulas, which in turn give new formulas, which in turn spawn new labels, from which we obtain new formulas, etc. This means that in its full generality the language of dynamic logic is highly nested and a formula can have a rather intricate structural complexity. Consequently some extra care must be taken when working through a proof by induction on the structure of a formula (which, in practice, means almost all proofs).

In this book we will not concern ourselves with this second problem. We restrict our attention to the first problem, namely of how the *-closure of a relation can be captured in modal logic. If you are interested in a fuller development of dynamic logic you may like to start with [27, Chapter 10] and then move on to [28].

14.2   The *-closure of a relation

Given a relation $\longrightarrow$ on a set A, the *-*closure* of $\longrightarrow$ is the least reflexive and transitive relation on A which includes $\longrightarrow$. Thus if $\overset{*}{\longrightarrow}$ is this *-closure, then

$$a \overset{*}{\longrightarrow} b$$

holds precisely when there is a chain $x_0, x_1, \ldots, x_n$ of elements of A with

$$a = x_0 \longrightarrow x_1 \longrightarrow \cdots \longrightarrow x_n = b.$$

In particular, the case $n = 0$ ensures that $\overset{*}{\longrightarrow}$ is reflexive, and the case $n = 1$ ensures that $\overset{*}{\longrightarrow}$ includes $\longrightarrow$. Since such chains can be concatenated, we see that $\overset{*}{\longrightarrow}$ is transitive.

In this chapter we consider a modal language with just two labels and corresponding modal operators $\square$ and $[\bullet]$. Thus the transition structures for this language have the form

$$\mathcal{A} = (A, \longrightarrow, \overset{\bullet}{\longrightarrow}).$$

We say such a structure is a *-*structure* if $\overset{\bullet}{\longrightarrow}$ is the *-closure of $\longrightarrow$. We will exhibit a set of axioms which ensure that a structure (of the above signature) is a *-structure, and discuss the completeness of the corresponding formal system.

14.3   The axioms for SLL

Consider the set of all formulas of the following shapes.

$$(*1) \quad [\bullet]\phi \to \phi$$
$$(*2) \quad [\bullet]\phi \to \square\,[\bullet]\phi$$
$$(*3) \quad [\bullet](\phi \to \square\phi) \to (\phi \to [\bullet]\phi)$$

These axioms, which are sometimes known as 'Segerberg's axioms', are an important part of the axioms for dynamic logic, and although they are still somewhat weaker than this full set, they do provide an example of a formal system which is slightly more interesting than most of the ones considered so far in this book. We name this standard formal system SLL.

Note that both $(*1)$ and $(*2)$ are general confluence axioms (as described in Chapter 6). In particular we know that $(*1)$ ensures that the relation $\overset{\bullet}{\longrightarrow}$ is reflexive, and $(*2)$ ensures that the composite $\longrightarrow\overset{\bullet}{\longrightarrow}$ is included $\overset{\bullet}{\longrightarrow}$. The shape $(*3)$ is new, and has the general form of an induction axiom.

The combined content of $(*1, 2, 3)$ is given by the following correspondence result.

14.1 THEOREM. *For each structure*

$$\mathcal{A} = (A, \longrightarrow, \stackrel{\bullet}{\longrightarrow})$$

the conditions

(i) $\mathcal{A}$ *is a ∗-structure*
(ii) $\mathcal{A}$ *models* SLL
(iii) *For some variable P, the structure* $\mathcal{A}$ *models the three formulas*

$$[\bullet]P \to P$$
$$[\bullet]P \to \Box\,[\bullet]P$$
$$[\bullet](P \to \Box P) \to (P \to [\bullet]P).$$

are equivalent.

Proof. (i) ⇒ (ii), Suppose that $\mathcal{A}$ is a ∗-structure. From the confluence results we know that $\mathcal{A}$ models (∗1) and (∗2), so it remains to check (∗3).

Suppose, for some formula ϕ, valuation α, and element a, that

$$a \Vdash [\bullet](\phi \to \Box\phi) \quad , \quad a \Vdash \phi$$

and consider any element b with

$$a \stackrel{\bullet}{\longrightarrow} b.$$

We must show that $b \Vdash \phi$. By (i) there is a chain of elements

$$a = x_0 \longrightarrow x_1 \longrightarrow \cdots \longrightarrow x_n = b.$$

Note that $a \stackrel{\bullet}{\longrightarrow} x_r$ for each $0 \le r \le n$ and hence

$$x_r \Vdash \phi \to \Box\phi.$$

This gives

$$x_r \Vdash \phi \Rightarrow x_{r+1} \Vdash \phi$$

so that (since $x_0 \Vdash \phi$) we have $x_n \Vdash \phi$, i.e. $b \Vdash \phi$, as required.

(ii) ⇒ (iii). This is trivial.

(iii) ⇒ (i). Suppose that (iii) holds. From the first two components we know that

$$\stackrel{\bullet}{\longrightarrow} \text{ is reflexive}$$

and

$$a \longrightarrow b \stackrel{\bullet}{\longrightarrow} c \Rightarrow a \stackrel{\bullet}{\longrightarrow} c$$

(for all $a, b, c \in A$). In particular we may set $b = c$ to get

$$a \longrightarrow b \Rightarrow a \stackrel{\bullet}{\longrightarrow} b$$

i.e. $\longrightarrow$ is included in $\overset{\bullet}{\longrightarrow}$. But now we have

$$a \longrightarrow b \longrightarrow c \;\Rightarrow\; a \longrightarrow b \overset{\bullet}{\longrightarrow} c \;\Rightarrow\; a \overset{\bullet}{\longrightarrow} c$$

and then a simple induction shows that the $*$-closure $\overset{*}{\longrightarrow}$ is included in $\overset{\bullet}{\longrightarrow}$. It thus remains to verify the converse.

For a fixed element a, consider any valuation α on $\mathcal{A}$ such that for all $x \in A$

$$x \Vdash P \;\Rightarrow\; a \overset{*}{\longrightarrow} x$$

(where P is the given variable of (iii)). Consider any element b with

$$a \overset{\bullet}{\longrightarrow} b \quad , \quad b \Vdash P.$$

Then $a \overset{*}{\longrightarrow} b$ so that, for each element x

$$b \longrightarrow x \;\Rightarrow\; a \overset{*}{\longrightarrow} b \overset{*}{\longrightarrow} x \;\Rightarrow\; a \overset{*}{\longrightarrow} x \;\Rightarrow\; x \Vdash P$$

i.e. $b \Vdash \Box P$. Thus

$$a \Vdash [\bullet](P \to \Box P),$$

and, trivially, $a \Vdash P$. The third component of the hypothesis (iii) now gives

$$a \Vdash [\bullet]P$$

so that

$$a \overset{\bullet}{\longrightarrow} b \;\Rightarrow\; a \overset{*}{\longrightarrow} b$$

as required. ∎

Now that we have this correspondence result it is tempting to try to follow the usual path to show that SLL is Kripke-complete. However, this path doesn't reach the required destination.

14.4 SLL is not canonical

As with every standard formal system, SLL has a canonical valued model $(\mathfrak{S}, \sigma)$ which is characteristic for SLL (in the sense that

$$\vdash_{\mathsf{SLL}} \phi \;\Leftrightarrow\; (\mathfrak{S}, \sigma) \Vdash^v \phi$$

for all formulas ϕ). However, unlike the other systems we have considered so far the unadorned structure $\mathfrak{S}$ does not model SLL, i.e. the system is not canonical.

To fix the notation let

$$\mathfrak{S} = (\boldsymbol{S}, \longrightarrow, \overset{\bullet}{\longrightarrow}).$$

This structure $\mathfrak{S}$ does have some of the properties required to be a $*$-structure.

14.2   LEMMA. *(a) The relation* $\overset{\bullet}{\longrightarrow}$ *is reflexive and transitive.*
(b) The relation $\longrightarrow$ *is included in* $\overset{\bullet}{\longrightarrow}$.

Proof. (a) As before, axiom (∗1) ensures that $\overset{\bullet}{\longrightarrow}$ is reflexive. Note also
that a combination of axioms (∗2, 3) and rule (N) gives

$$\vdash_{\mathsf{SLL}} \quad [\bullet]\phi \to [\bullet]^2\phi$$

and this ensures that $\overset{\bullet}{\longrightarrow}$ is transitive.

(b) Consider any $s, t \in S$ with $s \longrightarrow t$. Then, for each formula ϕ, axioms
(∗1, 2) give

$$[\bullet]\phi \in s \;\Rightarrow\; \square\,[\bullet]\phi \in s \;\Rightarrow\; [\bullet]\phi \in t \;\Rightarrow\; \phi \in t$$

so that $s \overset{\bullet}{\longrightarrow} t$, as required. ∎

An immediate consequence of this result is that the ∗-closure $\overset{\bullet}{\longrightarrow}$ of $\longrightarrow$
is included in $\overset{\bullet}{\longrightarrow}$. We will see how the converse fails, and hence $\mathfrak{S}$ is not a
∗-structure. To do this consider the ∗-structure

$$\mathcal{N} \;=\; (\mathsf{N}, \longrightarrow, \overset{\bullet}{\longrightarrow})$$

where for each $m, n \in \mathsf{N}$

$$m \longrightarrow n \;\Leftrightarrow\; n = m + 1.$$

Since, by definition, $\overset{\bullet}{\longrightarrow}$ is the ∗-closure of $\longrightarrow$, we have

$$m \overset{\bullet}{\longrightarrow} n \;\Leftrightarrow\; m \leq n.$$

Trivially $\mathcal{N}$ models **SLL**.

For a fixed $n \in \mathsf{N}$ and a variable P, consider any valuation on $\mathcal{N}$ for which

$$0, 1, \ldots, n \Vdash P \quad , \quad n + 1 \Vdash \neg P.$$

Thus

$$0 \Vdash P \wedge \square P \wedge \cdots \wedge \square^n P \wedge \neg\,[\bullet]P.$$

This shows that each finite part of

$$\Phi \;=\; \{\,\square^k P \mid k \in \mathsf{N}\} \cup \{\neg\,[\bullet]P\}$$

has a pointed valued model, and hence is **SLL**-consistent.

14.3   THEOREM. *The canonical structure* $\mathfrak{S}$ *is not a* ∗-*structure.*

Proof. The SLL-consistent set Φ exhibited above provides us with some $s \in \mathbf{S}$ with

$$\Box^n P \in s \quad \text{and} \quad \neg [\bullet] P \in s$$

for all $n \in \mathbb{N}$. The negated formula gives us some $t \in \mathbf{S}$ with

$$s \xrightarrow{\bullet} t \quad , \quad \neg P \in t.$$

But then $s \xrightarrow{*} t$ can not hold, for otherwise there is some sequence

$$s = s_0 \longrightarrow s_1 \longrightarrow \cdots \longrightarrow s_k = t$$

which (since $\Box^k P \in s$) gives $P \in t$.

This shows that $\xrightarrow{\bullet}$ is not included in $\xrightarrow{*}$ which is enough for the required result. ∎

Theorem 14.3 shows that SLL is not canonical but, as we will see in the next section, it is still Kripke-complete and decidable.

14.5 A filtration construction

In this section we look at a construction which will provide the basis of a proof that SLL has the fmp (and hence is Kripke-complete and decidable). Thus, consider any structure

$$\mathcal{A} = (A, \longrightarrow, \xrightarrow{\bullet})$$

together with a valuation α on $\mathcal{A}$ for which

$$(\mathcal{A}, \alpha) \text{ models SLL.} \tag{14.1}$$

In the eventual application of the construction this structure will be given to us (and in all cases could be taken to be the canonical valued structure for SLL). Note that we are not assuming that $\mathcal{A}$ is a *-structure.

As usual the construction also depends on a given set of formulas Γ. This set will have the following closure properties.

$$\Gamma \text{ is closed under subformulas.} \tag{14.2}$$

$$\text{For each formula } \theta, \quad [\bullet]\theta \in \Gamma \;\Rightarrow\; \Box\,[\bullet]\theta \in \Gamma. \tag{14.3}$$

$$\Gamma \text{ is finite.} \tag{14.4}$$

Note that for any formula ϕ there is at least one set Γ satisfying (14.2,14.3,14.4) with $\phi \in \Gamma$. In fact there is a smallest such Γ obtained by first closing off under subformulas and then closing off under property (14.3). The fact that this Γ is finite is something that you should verify.

With this structure $\mathcal{A}$ and set Γ we consider the usual equivalence relation $\sim$ on A given by

$$a \sim x \iff (\forall \theta \in \Gamma)[a \Vdash \theta \iff x \Vdash \theta]$$

(for $a, x \in A$). We then set

$$B = A/\!\!\sim$$

and we let $a \mapsto a^\sim$ be the canonical quotient function $A \longrightarrow B$. Note that, by (14.4), the set B is finite. We will construct a certain valued structure $(\mathcal{B}, \beta)$ based on B.

We make use of a definability property (which is reminiscent of an earlier result).

14.4 LEMMA. *For each element a of A and subset Y of B, there are formulas λ_a and μ_Y such that*

(l) $x \sim a \iff x \Vdash \lambda_a$
(m) $x^\sim \in Y \iff x \Vdash \mu_Y$

for each $x \in A$.

Proof. We first set

$$\lambda_a = \bigwedge\{\theta \in \Gamma \mid a \Vdash \theta\} \wedge \neg \bigvee\{\theta \in \Gamma \mid a \Vdash \neg\theta\}$$

and then set

$$\mu_Y = \lambda_{a(1)} \vee \cdots \vee \lambda_{a(n)}$$

where $a(1)^\sim, \ldots, a(n)^\sim$ is an enumeration of Y. ∎

Next let $\overset{\approx}{\longrightarrow}$ be the relation on B given by

$$b \overset{\approx}{\longrightarrow} y \iff (\exists a \in b, x \in y)[a \longrightarrow x]$$

(for $b, y \in B$), let $\overset{*}{\longrightarrow}$ be the $*$-closure of $\overset{\approx}{\longrightarrow}$, and set

$$\mathcal{B} = (B, \overset{\approx}{\longrightarrow}, \overset{*}{\longrightarrow}).$$

By construction $\mathcal{B}$ is a $*$-structure and hence is a model of **SLL**. Lemma 14.4 gives us some information about the definability of $\overset{*}{\longrightarrow}$.

14.5 LEMMA. *For each element a of $\mathcal{A}$ there is a formula ν_a such that for all $x \in A$*

$$a^\sim \overset{*}{\longrightarrow} x^\sim \iff (\mathcal{A}, \alpha, x) \Vdash \nu_a.$$

Furthermore, we have $(\mathcal{A}, \alpha, a) \Vdash [\bullet]\nu_a$.

Proof. For a given $a \in A$ let

$$Y = \{x^{\sim} \mid x \in A \text{ and } a^{\sim} \xrightarrow{\ *\ } x^{\sim}\}$$

and set $\nu_a = \mu_Y$. The required equivalence follows by part (m) of Lemma 14.4. For the second part we first show that

$$(\mathcal{A}, \alpha, a) \Vdash [\bullet](\nu_a \to \Box \nu_a). \tag{14.5}$$

To this end consider any element x of $\mathcal{A}$ with

$$a \xrightarrow{\ \bullet\ } x \quad , \quad x \Vdash \nu_a.$$

This second condition gives

$$a^{\sim} \xrightarrow{\ *\ } x^{\sim}$$

and we require $x \Vdash \Box \nu_a$. Thus consider any element u of $\mathcal{A}$ with $x \longrightarrow u$. Then, by construction of $\xrightarrow{\approx}$, we have

$$a^{\sim} \xrightarrow{\ *\ } x^{\sim} \xrightarrow{\approx} u^{\sim}$$

so that

$$a^{\sim} \xrightarrow{\ *\ } u^{\sim}$$

and hence $u \Vdash \nu_a$, as required.

This verifies (14.5).

Now, by (14.1), we know that $(\mathcal{A}, \alpha)$ models the axiom $(*3)$, so that

$$(\mathcal{A}, \alpha, a) \Vdash \nu_a \to [\bullet]\nu_a.$$

But, trivially, $a \Vdash \nu_a$, and hence $a \Vdash [\bullet]\nu_a$, as required. ∎

It is worth observing that the full power of (14.1) has not yet been used. So far we have used only that $(\mathcal{A}, \alpha)$ models $(*3)$; as yet the axioms $(*1, 2)$ have not been needed.

14.6 COROLLARY. *The assignment $a \mapsto a^{\sim}$ provides a surjective morphism*

$$\mathcal{A} \longrightarrow \mathcal{B}$$

that is, the implications

$$a \longrightarrow x \;\Rightarrow\; a^{\sim} \xrightarrow{\approx} x^{\sim}$$
$$a \xrightarrow{\ \bullet\ } x \;\Rightarrow\; a^{\sim} \xrightarrow{\ *\ } x^{\sim}$$

hold for all $a, x \in A$.

Proof. The first implication holds by construction and the second holds since $a \Vdash [\bullet]\nu_a$. The surjectivity is trivial. ∎

Of course we are working towards something stronger than this.

Notice that, by construction of $\sim$, for each variable $P \in \Gamma$ and $b \in B$ we have

$$(\exists a \in b)[a \Vdash P] \Leftrightarrow (\forall a \in b)[a \Vdash P].$$

Thus we may define a valuation β on B such that for all $P \in \Gamma$ and $a \in A$ we have

$$a \in \alpha(P) \Leftrightarrow a^{\sim} \in \beta(P)$$

with the values $\beta(P)$ for other P irrelevant. We fix such a valuation on B.

14.7 LEMMA. *A Γ-filtration*

$$(\mathcal{A}, \alpha) \longrightarrow (\mathcal{B}, \beta)$$

is provided by the assignment $a \mapsto a^{\sim}$.

Proof. From the above remarks it suffices to show the following for all elements a, x of $\mathcal{A}$ and all formulas θ.

- If $a^{\sim} \xrightarrow{\approx} x^{\sim}$ and $\Box\theta \in \Gamma$,   then $a \Vdash \Box\theta \Rightarrow x \Vdash \theta$.

- If $a^{\sim} \xrightarrow{*} x^{\sim}$ and $[\bullet]\theta \in \Gamma$,   then $a \Vdash [\bullet]\theta \Rightarrow x \Vdash \theta$.

The first of these holds by the construction of $\xrightarrow{\approx}$ from $\longrightarrow$.

For the second, suppose that

$$a^{\sim} \xrightarrow{*} x^{\sim}$$

so there is a sequence of transitions

$$a^{\sim} = a_0^{\sim} \xrightarrow{\approx} a_1^{\sim} \xrightarrow{\approx} \cdots \xrightarrow{\approx} a_n^{\sim} = x^{\sim}.$$

For any formula θ with $[\bullet]\theta \in \Gamma$ we have $\Box[\bullet]\theta \in \Gamma$ (by closure property (14.3)), and hence, for each $0 \le r < n$, a use of axiom ($*2$) gives

$$a_r \Vdash [\bullet]\theta \Rightarrow a_r \Vdash \Box[\bullet]\theta \Rightarrow a_{r+1} \Vdash [\bullet]\theta$$

which, by a simple induction, gives

$$a \Vdash [\bullet]\theta \Rightarrow x \Vdash [\bullet]\theta.$$

The required result now follows by axiom ($*1$). ∎

14.6   The completeness result

Let $\mathbb{M}$ be the class of all $*$-structures and let $\mathbb{M}(fin)$ be the class of all finite $*$-structures. Both of these are classes of models of SLL and, in fact, characteristic classes.

14.8   THEOREM. *For each formula ϕ, the three conditions*

   (i) $\vdash_{\mathsf{SLL}} \phi$
   (ii) $\mathbb{M}$ *models* ϕ
   (iii) $\mathbb{M}(fin)$ *models* ϕ

are equivalent.

Proof. Only the implication (iii) $\Rightarrow$ (i) is non-trivial.
Suppose that $\mathbb{M}(fin)$ models ϕ and consider any valued structure $(\mathcal{A}, \alpha)$ which is characteristic for SLL (say, the canonical valued model). We show that $(\mathcal{A}, \alpha) \Vdash^v \phi$.
Consider any set of formulas Γ satisfying (14.2,14.3,14.4) with $\phi \in \Gamma$, and let $(\mathcal{B}, \beta)$ be the Γ-filtration of $(\mathcal{A}, \alpha)$ constructed in the previous section. Then

$$(\mathcal{A}, \alpha) \Vdash^v \psi \iff (\mathcal{B}, \beta) \Vdash^v \psi$$

for all formulas $\psi \in \Gamma$. In particular, since $\mathcal{B}$ is a finite model of SLL, we have, $\mathcal{B} \Vdash^u \phi$, so that $(\mathcal{B}, \beta) \Vdash^v \phi$, and hence $(\mathcal{A}, \alpha) \Vdash^v \phi$ which gives the required result. ∎

This result completes the proof of the following.

14.9   THEOREM. *The formal system SLL has the fmp (and hence is Kripke-complete) but is not canonical.*

14.7   Exercises

14.1   Show directly that $\vdash_{\mathsf{SLL}} [\bullet]\phi \to [\bullet]^2\phi$.

14.2   Consider a ticking clock structure $\mathcal{A}$ as in Exercise 8.14 of Chapter 8, with next instance function next. This gives a model of Segerberg's axioms (with $\square$ replaced by $\bigcirc$). Let S be the formal system axiomatized by this form of Segerberg's axioms together with the shape

$$\bigcirc\phi \leftrightarrow \neg \bigcirc \neg\phi.$$

Given an element a of $\mathcal{A}$ and $r \in \mathsf{N}$, write

$$a(r) \quad \text{for} \quad \mathsf{next}^r(a).$$

Given a valuation α on $\mathcal{A}$ let ν be the valuation on $\mathcal{N}$ (the structure used in Section 14.4) given by

$$r \in \nu(P) \iff a(r) \in \alpha(P)$$

(for all variables P).

(a) Show that the assignment $r \mapsto a(r)$ is a zigzag morphism

$$(\mathcal{N}, \nu) \longrightarrow (\mathcal{A}, \alpha).$$

(b) Show that both

$$(\mathcal{N}, \nu, r) \Vdash^p \phi \iff (\mathcal{A}, \alpha, a(r)) \Vdash^p \phi$$

and

$$\mathcal{N} \Vdash^u \phi \implies \mathcal{A} \Vdash^u \phi$$

hold for all $r \in \mathbb{N}$ and formulas ϕ.

(c) Find an example to show that this second implication need not be an equivalence.

(d) Show that S is not canonical.

14.3   For $m, n \in \mathbb{N}$ an (m, n)-spoon is a set A of $m + n + 1$ elements

$$a = a_0, a_1, \ldots, a_m = c \quad , \quad c = b_0, b_1, \ldots, b_n = c$$

furnished with a 'next' function given by

$$\begin{aligned} \mathsf{next}(a_i) &= a_{i+1} && \text{for } 0 \le i < m \\ \mathsf{next}(b_j) &= b_{j+1} && \text{for } 0 \le j < n. \end{aligned}$$

Pictorially this is

This gives us a ticking clock structure $\mathcal{A} = (A, \longrightarrow, \overset{\bullet}{\longrightarrow})$ which eventually cycles and which, of course, models SLL.

Let

$$f : \mathsf{N} \longrightarrow A$$

be the function given by

$$
\begin{aligned}
f(i) &= a_i & \text{for } 0 \leq i < m \\
f(m + kn + j) &= b_j & \text{for } 0 \leq j < n
\end{aligned}
$$

(in particular $f(m) = c$).

It can be shown that for each formula ϕ and valuation ν on N with

$$(\mathcal{N}, \nu, 0) \Vdash \phi$$

there are $m, n \in \mathsf{N}$ and a valuation μ on N such that

$$(\mathcal{N}, \mu, 0) \Vdash \phi$$

and

$$m + r \in \mu(P) \iff m + n + r \in \mu(P)$$

for all $r \in \mathsf{N}$ and variables P in ϕ.

(a) Show that

$$\mathsf{next}^s(f(r)) = f(r + s)$$

for all $r, s \in \mathsf{N}$.

(b) Show that f is a p-morphism $\mathcal{N} \longrightarrow \mathcal{A}$.

(c) Show that for each formula ϕ,

$$\mathcal{N} \Vdash^u \phi$$

holds if and only if

$$\mathcal{A} \Vdash^u \phi$$

holds for all spoons $\mathcal{A}$.

14.4 For an arbitrary set A let $\square$ be an arbitrary box operator on $\mathcal{P}A$ (where $\square$ need not be induced by a transition relation on A). An *S-companion* of $\square$ is a box operator $[\bullet]$ on $\mathcal{P}A$ satisfying

(∗1) $[\bullet]X \subseteq X$

(∗2) $[\bullet]X \subseteq \square [\bullet]X$

(∗3) $[\bullet](X \to \square X) \subseteq (X \to [\bullet]X)$

for all $X \subseteq A$. (Thus the pair $\square$, $[\bullet]$ satisfy Segerberg's axioms).
    For each $X \subseteq A$ set
$$DX = X \cap \square X.$$
and let
$$(D^\alpha X \mid \alpha \in Ord)$$
be the ordinal indexed chain defined by
$$\begin{aligned}
D^0 X &= X \\
D^{\alpha+1} &= D(D^\alpha X) \\
D^\lambda &= \cap\{D^\alpha X \mid \alpha < \lambda\}
\end{aligned}$$

(for each ordinal α and limit ordinal λ).

(a) Show that an S-companion is idempotent.

(b) Show that $\square$ has at most one S-companion.

(c) Show that D is deflationary, monotone, and $\cap$-preserving.

(d) Show that for each $X \subseteq A$ there is a unique largest $Y \subseteq X$ with $DY = Y$.

(e) Show that setting
$$[\bullet]X = \text{the } Y \text{ of (d)}$$
produces the S-companion of $\square$.

(f) Show that
$$[\bullet]X = D^\infty X$$
for some suitably large ordinal ∞.

(g) Show that if $\square$ is induced by a transition relation then this ordinal ∞ is just ω.

14.5 Suppose that $\mathcal{A}$ is a model of SLL and the generating transition relation is already reflexive.

(a) Show that $\mathcal{A}$ models
$$[\bullet]\Diamond P \rightarrow \langle\bullet\rangle \square P$$
if and only if for each element a there is some element b with $a \overset{\bullet}{\longrightarrow} b$ and where b can see only itself.

(b) Does this generalize the result of Exercise 5.8 of Chapter 5?

Chapter 15

Löb logic

15.1 Introduction

Löb logic is the study of the shape $L(\phi)$, i.e.

$$\Box(\Box\phi \to \phi) \to \Box\phi.$$

We have seen that the models of this shape are the transitive well-founded structures, but apart from that, you may ask, what is so special about it? Why pick out this one from amongst a whole host of equally dreary looking shapes? To answer this question, and hence motivate the study of Löb logic, we must look at the history of an apparently unconnected part of mathematics; namely Gödel's incompleteness theorem.

Consider the first order structure

$$\mathcal{N} = (\mathsf{N}, +, \times, \leq, 0, 1)$$

that is, the natural numbers N furnished with the usual arithmetical attributes. Elementary arithmetic is the study of the first order properties of $\mathcal{N}$. To do this we work in an associated first order language $\mathcal{L}$ and analyse various recursively axiomatizable theories T which, we hope, will completely capture the elementary properties of $\mathcal{N}$. The main theory studied is Peano Arithmetic, but there are others as well.

By construction (or assumption), the theory T is sound in the sense that

$$T \vdash \sigma \ \Rightarrow\ \mathcal{N} \models \sigma \tag{15.1}$$

for each sentence σ of $\mathcal{L}$. (Here, $\vdash$ and $\models$ are the first order entailment and satisfaction relations. They should not be confused with modal relations with the same notation.)

The objective of the whole exercise is to obtain a recursively axiomatized theory T for which the soundness implication (15.1) is an equivalence. Before 1930 it was more or less expected that eventually Peano Arithmetic would turn

201

202 CHAPTER 15. LÖB LOGIC

out to be such a theory, but then Gödel proved his incompleteness theorem which simply says that there is no such theory. The method of proof used by Gödel is to show how the external behaviour of T and its associated entailment relation $\vdash$ can be mimicked within T. Thus we first show how each sentence σ has a name $\ulcorner \sigma \urcorner$ in $\mathcal{L}$ which is a numeral (denoting some natural number and called the gödel number of σ). We then construct a certain formula $Th(\cdot)$ of $\mathcal{L}$ (containing just one free variable) which captures $\vdash$ in the sense that for each sentence σ,

$$T \vdash \sigma \Leftrightarrow \mathcal{N} \models Th(\ulcorner \sigma \urcorner). \qquad (15.2)$$

Here $Th(\ulcorner \sigma \urcorner)$ is the sentence obtained by replacing each occurrence of the free variable in $Th(\cdot)$) by the numeral $\ulcorner \sigma \urcorner$.

The next step is to show how many of the important consequences of (15.2) are actually deducible in T. It turns out that these consequences can be reduced to just three instances, called the *derivability conditions*. (They were first isolated by Hilbert and Bernays, and then slightly modified by Löb.) The conditions assert that for all sentences σ and τ,

(D1) $T \vdash \sigma \Rightarrow T \vdash Th(\ulcorner \sigma \urcorner)$
(D2) $T \vdash Th(\ulcorner \sigma \to \tau \urcorner). \to . Th(\ulcorner \sigma \urcorner) \to Th(\ulcorner \tau \urcorner)$
(D3) $T \vdash Th(\ulcorner \sigma \urcorner) \to Th(\ulcorner Th(\ulcorner \sigma \urcorner) \urcorner)$

hold. Notice how these are reminiscent of the rule(N) and the axioms K and 4 of modal logic.

The final component in the Gödel incompleteness proof is the fixed point property. In the first instance this asserts that for each formula $\phi(\cdot)$ with just one free variable, there is a sentence δ with

$$T \vdash \delta \leftrightarrow \phi(\ulcorner \delta \urcorner).$$

For most purposes we don't need this generality; the following instance will suffice.

(Δ) For each sentence σ there is some sentence δ such that

$$T \vdash \delta \leftrightarrow [Th(\ulcorner \delta \urcorner) \to \sigma]$$

holds.

Notice that if we take the trivially false sentence $\perp$ for σ, then (Δ) provides us with a sentence γ with

$$T \vdash \gamma \leftrightarrow \neg Th(\ulcorner \gamma \urcorner). \qquad (15.3)$$

By (15.1) this displayed equivalence holds in $\mathcal{N}$ (i.e. is true) and then (15.2) provides us with an informal reading of γ,

γ: I am not provable (in T).

A few more mental acrobatics shows that this sentence γ ought to be true, and hence

$$\neg[T \vdash \gamma] \quad , \quad \mathcal{N} \models \gamma$$

so that the implication (15.1) can not be reversed. In detail, using (15.3) and (15.1) and then (15.2) and finally (15.1) again, we have

$$\mathcal{N} \models \neg\gamma \; \Rightarrow \; \mathcal{N} \models Th(\ulcorner\gamma\urcorner) \; \Rightarrow \; T \vdash \gamma \; \Rightarrow \; \mathcal{N} \models \gamma$$

and hence $\mathcal{N} \models \gamma$. But then $\mathcal{N} \models \neg Th(\ulcorner\gamma\urcorner)$ and hence $\neg[T \vdash \gamma]$ (by (15.3) and (15.2)), as required.

The informal reading of the Gödel sentence γ suggests that we should also look at a formalized version of

I am not provable

i.e. a sentence η with

$$T \vdash \eta \leftrightarrow Th(\ulcorner\eta\urcorner). \tag{15.4}$$

(The fixed point property ensures that such sentences do exist, although they are not immediately obtainable from the version (Δ).)

Twenty years after Gödel's result (i.e. around 1950) Henkin asked about the truth value of η. Unlike γ, this is not immediately decided by informal arguments. The correct answer was given a few years later by Löb, and this is where the derivability conditions come into their own.

Given an arbitrary sentence η satisfying (15.4), use (Δ) to obtain a sentence δ such that

$$T \vdash \delta \leftrightarrow [Th(\ulcorner\delta\urcorner) \rightarrow \eta].$$

Using the implication $\rightarrow$, applications of $(D1, 2)$ give

$$T \vdash Th(\ulcorner\delta\urcorner) \rightarrow Th(\ulcorner\sigma\urcorner)$$

and

$$T \vdash Th(\ulcorner\sigma\urcorner) \rightarrow .Th(\ulcorner Th(\ulcorner\delta\urcorner)\urcorner) \rightarrow Th(\ulcorner\eta\urcorner)$$

where here

$$\sigma \text{ is } Th(\ulcorner\delta\urcorner) \rightarrow \eta.$$

But now $(D3)$ gives

$$T \vdash Th(\ulcorner\delta\urcorner) \rightarrow Th(\ulcorner\eta\urcorner)$$

which, by (15.4), gives

$$T \vdash Th(\ulcorner\delta\urcorner) \rightarrow \eta$$

and hence we have verified that $T \vdash \delta$. A final application of $(D1)$ gives $T \vdash Th(\ulcorner\delta\urcorner)$ which, from above, shows that $T \vdash \eta$. Thus the Henkin sentence is provable (and hence true).

It is clear from this argument that once we have used (Δ) to obtain the sentence δ, the rest is a modal computation in the system **K4**. This prompts us to look at whether **K4** can be extended so as to incorporate a suitable version of (Δ). The crucial observation was made by Macintyre and Simmons in [39] (although there it is not expressed in modal terms).

15.1 THEOREM. *Let* S *be any standard formal system extending* **K4**. *For each formula ϕ, the following are equivalent.*

(i) $\vdash_S L(\phi)$.
(ii) There is a formula δ such that $\vdash_S \delta \leftrightarrow (\Box \delta \rightarrow \phi)$.

Furthermore, if there is such a formula δ then

$$\vdash_S \Box\delta \leftrightarrow \Box\phi \quad , \quad \vdash_S \delta \leftrightarrow (\Box\phi \rightarrow \phi)$$

and hence δ is essentially unique.

Proof. (i) $\Rightarrow$ (ii). Suppose that (i) holds, i.e. that

$$\vdash_S \Box(\Box\phi \rightarrow \phi) \rightarrow \Box\phi$$

and set

$$\delta := \Box\phi \rightarrow \phi$$

so that

$$\vdash_S \Box\delta \rightarrow \Box\phi.$$

But, trivially,

$$\vdash_S \phi \rightarrow \delta$$

so that, by (EN), we have

$$\vdash_S \Box\phi \rightarrow \Box\delta.$$

Thus

$$\vdash_S \Box\delta \leftrightarrow \Box\phi$$

and we may replace $\Box\phi$ by $\Box\delta$ to get

$$\vdash_S \delta \leftrightarrow (\Box\delta \rightarrow \phi)$$

as required.
(ii) $\Rightarrow$ (i). Conversely, suppose that

$$\vdash_S \delta \leftrightarrow (\Box\delta \rightarrow \phi).$$

Then

$$\vdash_S \delta \wedge \Box\delta \rightarrow \phi \quad , \quad \vdash_S \phi \rightarrow \delta$$

so that (N) and K give us

$$\vdash_S \;\Box\delta \wedge \Box^2\delta \rightarrow \Box\phi \quad , \quad \vdash_S \;\Box\phi \rightarrow \Box\delta$$

which, using 4, shows that

$$\vdash_S \;\Box\delta \leftrightarrow \Box\phi$$

and hence

$$\vdash_S \;\delta \leftrightarrow (\Box\phi \rightarrow \phi).$$

This last equivalence gives

$$\vdash_S \;L(\phi) \leftrightarrow (\Box\delta \rightarrow \Box\phi)$$

so that, using the previous equivalence, we have $\vdash_S L(\phi)$, as required. ∎

This result opens up a whole new field of enquiry.

First of all the routine properties of the modal system **K4L** should be established. I do this in the rest of this chapter. In particular I show that **K4L** is not canonical but does have the fmp (and hence is Kripke-complete and decidable).

Theorem 15.1 shows how the shape L captures the fixed point principle (Δ). But what about other fixed point principles (such as the one required to obtain the Henkin sentence η)? In the early 1970s this question was studied intensively by two groups of people, one centred in Amsterdam and the other in Sienna. The outcome of this study was a quite general result of which a particular case is as follows.

Let ϕ be a modal formula containing a variable P all of whose occurrences lie within the scope of $\Box$. Then there is a formula δ such that

$$\vdash_{\mathsf{K4L}} \;\delta \leftrightarrow \phi[\delta \text{ for } P].$$

All variables in δ also appear in ϕ and, naturally, P does not occur in δ. There is a certain unicity about δ since

$$\vdash_{\mathsf{K4L}} \;(P \leftrightarrow \phi) \wedge \Box(P \leftrightarrow \psi). \rightarrow .(P \leftrightarrow \phi)$$

holds.

Another direction of enquiry concerns the interpretation of $\Box$ as $Th(\cdot)$, and to what extent the properties of $Th(\cdot)$ are captured by **K4L**. This was solved by Solovay around 1975 when he obtained a completeness result which, in imprecise terms, shows that **K4L** is exactly the theory of $Th(\cdot)$.

We won't look at these more advanced results in this book. If you are interested, a rather turgid account is given in [14]. A far more readable and comprehensive account is given in [44], and the survey article [43] is well worth a read. An amusing account of this use of modal logic is given in [45].

15.2   The system defined

Let LL be the standard formal system axiomatized by all formulas of the shape

$$L(\phi): \quad \Box(\Box\phi \to \phi) \to \Box\phi.$$

Note that we have not included the shape 4 as an axiom of LL. This is because we don't need to.

15.2   LEMMA. *For each formula* ϕ,

$$\vdash_{LL} \Box\phi \to \Box^2\phi$$

holds.

Proof. For the given formula ϕ set $\psi = \phi \wedge \Box\phi$. Then

$$\vdash_K \Box\psi \to \Box\phi \wedge \Box^2\phi$$

so that

$$\vdash_K \phi \to (\Box\psi \to \psi)$$

and hence

$$\vdash_K \Box\phi \to \Box(\Box\psi \to \psi).$$

Thus, using $L(\psi)$, we have

$$\vdash_{LL} \Box\phi \to \Box\psi$$

which, since $\vdash_K \Box\psi \to \Box^2\phi$, gives the required result. ∎

In Lemma 5.6 we saw that every model of LL is transitive and hence also models the shape 4. Lemma 15.2 is just a proof theoretic version of this result. This result shows that LL and K4L are essentially the same system.

In Theorem 5.7 we showed that the models of LL are precisely the well-founded, transitive structures. Later we will see that these structures characterize LL, but before we do that let us show that LL is not canonical.

15.3   The rule of disjunction

A formal system S is said to have the *rule of disjunction* if for all formulas $\phi_n, \ldots \phi_n$,

$$\vdash_S \Box\phi_1 \vee \cdots \vee \Box\phi_n \implies \vdash_S \phi_1 \text{ or } \cdots \text{ or } \vdash_S \phi_n.$$

In this section we see that LL has this rule and then use this to show that LL is not canonical.

Consider any valued structure $(\mathcal{A}, \alpha)$ which is characteristic for LL. i.e. such that

$$\vdash_{\mathsf{LL}} \phi \Leftrightarrow (\mathcal{A}, \alpha) \Vdash^v \phi$$

for all formulas ϕ. We modify $(\mathcal{A}, \alpha)$ to get a new structure. Thus let b be a new element (not in A) and set

$$B = A \cup \{b\}.$$

Let $\overset{*}{\longrightarrow}$ be the relation on B given by

$$x \overset{*}{\longrightarrow} y \Leftrightarrow \begin{cases} x, y \in A \quad \text{and} \quad x \longrightarrow y \\ \qquad\qquad \text{or} \\ x = b \quad \text{and} \quad y \in A \end{cases}$$

for each $x, y \in B$ (where $\longrightarrow$ is the distinguished relation of $\mathcal{A}$). Set

$$\mathcal{B} = (B, \overset{*}{\longrightarrow})$$

and let β be the valuation on $\mathcal{B}$ with

$$\beta(P) = \alpha(P)$$

for all variables P. (In particular $b \notin \beta(P)$.) It is an easy exercise to show that for each $a \in A$ and formula ϕ

$$(\mathcal{B}, \beta, a) \Vdash^p \phi \Leftrightarrow (\mathcal{A}, \alpha, a) \Vdash^p \phi \tag{15.5}$$

and hence

$$(\mathcal{B}, \beta, b) \Vdash^p \Box\phi \Leftrightarrow (\mathcal{A}, \alpha) \Vdash^v \phi \tag{15.6}$$

(where b is the distinguished element of $\mathcal{B}$).

15.3 LEMMA. *The valued structure* $(\mathcal{B}, \beta)$ *models* LL.

Proof. By (15.5) it suffices to show that

$$(\mathcal{B}, \beta, b) \Vdash^p \mathsf{L}(\phi)$$

for each formula ϕ. To this end suppose that

$$(\mathcal{B}, \beta, b) \Vdash^p \Box(\Box\phi \rightarrow \phi).$$

Then, using (15.6), we have

$$(\mathcal{A}, \alpha) \Vdash^v \Box\phi \rightarrow \phi$$

so that

$$(\mathcal{A}, \alpha) \Vdash^v \Box(\Box\phi \rightarrow \phi)$$

and hence, by L(ϕ),

$$(\mathcal{A}, \alpha) \Vdash^v \Box \phi.$$

These give

$$(\mathcal{A}, \alpha) \Vdash^v \phi$$

so that a second use of (15.6) gives

$$(\mathcal{B}, \beta, b) \Vdash^p \Box \phi$$

as required. ∎

Using this we quickly obtain the main result of this section.

15.4 THEOREM. *The system* **LL** *has the rule of disjunction.*

Proof. Suppose that

$$\vdash_{LL} \Box \phi_1 \vee \ldots \vee \Box \phi_n$$

for some formulas $\phi_1, \ldots, \phi_n$. Then, using Lemma 15.3 we have

$$(\mathcal{B}, \beta, b) \Vdash^p \Box \phi_1 \vee \ldots \vee \Box \phi_n$$

and hence

$$(\mathcal{B}, \beta, b) \Vdash^p \Box \phi_r$$

for some $1 \leq r \leq n$. This, with (15.6), gives

$$(\mathcal{A}, \alpha) \Vdash^v \phi_r$$

and hence $\vdash_{LL} \phi_r$, as required. ∎

Finally, for this section, we use the rule of disjunction to obtain some information about the canonical structure for **LL**. For this purpose let us say an element of a structure is *all seeing* if $a \longrightarrow x$ for all elements x of this structure. In particular such an element will be reflexive.

15.5 LEMMA. *The canonical structure* $\mathfrak{S}$ *of* **LL** *has an all seeing element.*

Proof. Consider the set of formulas

$$\Phi = \{\neg \Box \phi \mid \neg[\vdash_{LL} \phi]\}.$$

This is **LL**-consistent. For if not, there are formulas $\phi_1, \ldots, \phi_n$ with

$$\vdash_{LL} \neg \Box \phi_1 \wedge \cdots \wedge \neg \Box \phi_n \rightarrow \bot$$

i.e. with

$$\vdash_{LL} \Box \phi_1 \vee \cdots \vee \Box \phi_n.$$

The rule of disjunction now gives us some $1 \leq r \leq n$ with $\vdash_{LL} \phi_r$, which is clearly contradictory.

Since Φ is consistent, there is some $s \in S$ with $\Phi \subseteq s$. Then

$$\neg [\vdash_{LL} \phi] \Rightarrow \neg \square \phi \in s$$

for each formula ϕ, and hence for each $t \in S$

$$\square \phi \in s \Rightarrow \vdash_{LL} \phi \Rightarrow \phi \in t$$

so that $s \longrightarrow t$, and hence s is all seeing. ∎

Since all models of LL are well founded and in particular irreflexive, this immediately gives the following.

15.6 THEOREM. *The system* LL *is not canonical.*

15.4   The fmp

Let $\mathbb{M}$ be the set of all structures

$$\mathcal{B} = (B, \longrightarrow)$$

with the following properties.

(i) The carrier B is finite.

(ii) The relation $\longrightarrow$ is transitive and irreflexive.

(iii) The relation is tree-like in the sense that for each $x, y, z \in B$

$$\left. \begin{array}{c} x \longrightarrow z \\ y \longrightarrow z \end{array} \right\} \Rightarrow x \longrightarrow y \text{ or } x = y \text{ or } y \longrightarrow x.$$

(iv) There is a root, i.e. a (unique) element b such that $b \longrightarrow x$ for all other elements x.

For convenience let us refer to such structures as finite trees.

Condition (ii) ensure that any chain of elements

$$x_0 \longrightarrow x_1 \longrightarrow x_2 \longrightarrow \cdots$$

of a finite tree $\mathcal{B}$ can not contain repetitions hence, by (i), must be finite. Thus $\mathbb{M}$ is a class of models of LL. In this section we improve this observation.

15.7 THEOREM. *The class* $\mathbb{M}$ *of finite trees is characteristic for* LL, *i.e. for each formula ϕ the equivalence*

$$\vdash_{\mathsf{LL}} \phi \;\Leftrightarrow\; \mathbb{M} \text{ models } \phi$$

holds. In particular, LL *has the fmp and so is Kripke-complete and is decidable.*

Since $\mathbb{M}$ is a class of models of LL, soundness gives the implication $\Rightarrow$. The proof of the converse implication $\Leftarrow$ will take up the remainder of this section.

Fix some formula γ with $\neg[\vdash_{\mathsf{LL}} \gamma]$. We must produce some $\mathcal{B} \in \mathbb{M}$ with $\neg[\mathcal{B} \Vdash^u \gamma]$, that is, we must find some valuation β on $\mathcal{B}$ and an element b of $\mathcal{B}$ with

$$(\mathcal{B}, \beta, b) \Vdash^p \neg\gamma.$$

(The element b will turn out to be the root of $\mathcal{B}$.)

Let $(\mathcal{A}, \alpha)$ be any valued structure which is characteristic for LL (e.g. the canonical valued model). In particular we have $\neg[(\mathcal{A}, \alpha) \Vdash^v \gamma]$. As usual, $(\mathcal{B}, \beta)$ will be constructed from $(\mathcal{A}, \alpha)$. To do this we could use the filtration technique, but I have chosen not to do this so as to illustrate another technique. The elements of $\mathcal{B}$ will be certain finite chains of elements of $\mathcal{A}$ ordered by extension.

First of all we need to pick out a special element of $\mathcal{A}$.

15.8 LEMMA. *There is an element a of $\mathcal{A}$ such that*

$$(\mathcal{A}, \alpha, a) \Vdash^p \neg\gamma \wedge \square\gamma \tag{15.7}$$

holds.

Proof. By way of contradiction suppose there is no such element, so that

$$(\mathcal{A}, \alpha) \Vdash^v \square\gamma \to \gamma.$$

But then

$$(\mathcal{A}, \alpha) \Vdash^v \square(\square\gamma \to \gamma).$$

and hence axiom $\mathsf{L}(\gamma)$ gives

$$(\mathcal{A}, \alpha) \Vdash^v \square\gamma.$$

This, with the original hypothesis, gives

$$(\mathcal{A}, \alpha) \Vdash^v \gamma$$

which contradicts the chosen property of γ. ∎

We now fix an element with property (15.7). All our subsequent computations take place within $\mathcal{A}$.

Let Γ be the set of subformulas of γ. This set Γ is finite. The size of the constructed finite tree is determined by the cardinality $m =\mid \Gamma \mid$. (A crude upper bound for the size of the constructed $\mathcal{B}$ is m^m, however this can be considerably improved.)

To facilitate the construction of $\mathcal{B}$ we introduce some terminology.

We say an element x of $(\mathcal{A}, \alpha)$ is *relevant* if there is some formula $\phi \in \Gamma$ with

$$x \Vdash \neg \phi \wedge \square \phi.$$

In particular, the formula γ witnesses the relevance of the distinguished element a. A chain of elements

$$\mathsf{x} := x_1, x_2, \ldots, x_r, \ldots, x_n$$

of $\mathcal{A}$ is *acceptable* if each term x_r is relevant and

$$x_1 \longrightarrow x_2 \longrightarrow \cdots \longrightarrow x_r \longrightarrow \cdots \longrightarrow x_n.$$

Given such a chain x we write $\ell(\mathsf{x})$ for its last term x_n. Notice that there is at least one acceptable chain, namely the chain a of length 1 whose sole term is a (the distinguished element). Clearly $\ell(\mathsf{a}) = a$.

15.9 LEMMA. *Each acceptable chain has length no more than $\mid \Gamma \mid$.*

Proof. Consider any acceptable chain x (as above). Since each term x_r is relevant there is a sequence of formulas

$$\phi_1, \phi_2, \ldots, \phi_r, \ldots, \phi_n$$

where each is a member of Γ with

$$x_r \Vdash \neg \phi_r \wedge \square \phi_r$$

for $1 \leq r \leq n$. It suffices to show that these formulas are distinct, for then we have $n \leq \mid \Gamma \mid$, as required.

By way of contradiction, suppose there are $1 \leq r < s \leq n$ with

$$\phi_r = \phi_s = \phi \text{ (say)}.$$

Then, amongst other things

$$x_r \Vdash \square \phi \quad , \quad x_s \Vdash \neg \phi$$

which, since $x_r \longrightarrow x_s$, is the required contradiction. ∎

The elements of the required tree $\mathcal{B}$ will be certain acceptable chains x. The initial element x_1 of each selected x will be a and the relation of $\mathcal{B}$ will be the initial section ordering for chains.

We construct the carrying set B of $\mathcal{B}$ in layers

$$B_1, B_2, \ldots, B_r, \ldots, B_m$$

where $m = |\Gamma|$ and where each B_r consists of chains of length r.
The base and step constructions are as follows.

(1) B_1 has one member, namely the chain $\mathbf{a}$.

$(r \mapsto r + 1)$ Suppose we have constructed B_r with this set finite. Consider
all pairs $\mathbf{x}, \phi$ where

$$\mathbf{x} \in B_r \quad , \quad \phi \in \Gamma \quad , \quad \ell(\mathbf{x}) \Vdash \Diamond \neg \phi.$$

There are no more than $|B| \times |\Gamma|$ such pairs. (There may be no such
pairs, in which case $B_{r+1} = \emptyset$, and the construction terminates.) Note that
the axiom $L(\phi)$ gives

$$\ell(\mathbf{x}) \Vdash \Diamond(\neg \phi \wedge \Box \phi)$$

and hence there is some element y of $\mathcal{A}$ with $\ell(\mathbf{x}) \longrightarrow y$ and

$$y \Vdash \neg \phi \wedge \Box \phi.$$

In particular, the element y is relevant and the extended chain

$$\mathbf{y} = \mathbf{x} \frown y$$

is acceptable (with $\ell(\mathbf{y}) = y$). Select one such extension for each pair $\mathbf{x}, \phi$ and
let B_{r+1} be the set of selected extensions.

This completes the construction of the layers.
    Now set

$$B = B_1 \cup B_2 \cup \cdots \cup B_m$$

and let $\longrightarrow$ be the initial section ordering on B, i.e. for $x, y \in B$

$$\mathbf{x} \longrightarrow \mathbf{y} \Leftrightarrow \mathbf{x} \text{ is a proper initial section of } \mathbf{y}.$$

Let

$$\mathcal{B} = (B, \longrightarrow).$$

You should have no trouble verifying that $\mathcal{B} \in \mathbb{M}$. In particular, note that the
root of $\mathcal{B}$ is $\mathbf{a}$.

Observe also that the construction ensures the following.

(*) For each $\mathbf{x} \in B$ and $\phi \in \Gamma$ with

$$\ell(\mathbf{x}) \Vdash \Diamond \neg \phi$$

there is some $\mathbf{y} \in B$ with

$$\mathbf{x} \longrightarrow \mathbf{y} \quad , \quad \ell(\mathbf{y}) \Vdash \neg \phi \wedge \Box \phi.$$

In fact there is such a $\mathbf{y}$ with just one extra term.

Finally consider any valuation β on $\mathcal{B}$ such that

$$(\mathcal{B}, \beta, \mathsf{x}) \Vdash^p P \Leftrightarrow (\mathcal{A}, \alpha, \ell(\mathsf{x})) \Vdash^p P$$

for all variable $P \in \Gamma$. (The behaviour on other variables is irrelevant.) We prove the obvious extension of this equivalence.

15.10 LEMMA. *The equivalence*

$$(\mathcal{B}, \beta, \mathsf{x}) \Vdash^p \phi \Leftrightarrow (\mathcal{A}, \alpha, \ell(\mathsf{x})) \Vdash^p \phi$$

holds for all $\mathsf{x} \in B$ *and* $\phi \in \Gamma$.

Proof. We proceed by induction on the structure of ϕ. The base case holds by the choice of β, and the passage across propositional connectives is immediate. Thus it suffices to show that

$$\mathsf{x} \Vdash \Box\phi \Leftrightarrow \ell(\mathsf{x}) \Vdash \Box\phi$$

where $\phi \in \Gamma$ and we already have the corresponding equivalence for ϕ.

For the implication $\Rightarrow$, suppose that $\mathsf{x} \Vdash \Box\phi$ and, by way of contradiction, that

$$\ell(\mathsf{x}) \Vdash \neg \Box\phi.$$

Then, by $(*)$, there is some $\mathsf{y} \in B$ with

$$\mathsf{x} \longrightarrow \mathsf{y} \quad , \quad \ell(\mathsf{y}) \Vdash \neg\phi$$

and hence, by the Induction Hypothesis, $\mathsf{y} \Vdash \neg\phi$. This is the required contradiction.

For the implication $\Leftarrow$, suppose that $\ell(\mathsf{x}) \Vdash \Box\phi$ and consider any $\mathsf{y} \in B$ with $\mathsf{x} \longrightarrow \mathsf{y}$. Then, by construction of B,

$$\ell(\mathsf{x}) \longrightarrow \ell(\mathsf{y})$$

so that $\ell(\mathsf{y}) \Vdash \phi$ and the Induction Hypothesis gives $\mathsf{y} \Vdash \phi$, as required. ∎

As an application of this Lemma consider the case where $\mathsf{x} = \mathsf{a}$, the root of $\mathcal{B}$. Then $\ell(\mathsf{a}) = a$, the distinguished element of $\mathcal{A}$, and we have

$$(\mathcal{B}, \beta, \mathsf{a}) \Vdash^p \phi \Leftrightarrow (\mathcal{A}, \alpha, a) \Vdash^p \phi$$

for all $\phi \in \Gamma$. In particular, with $\phi = \gamma$, we have

$$(\mathcal{B}, \beta, \mathsf{a}) \Vdash \neg\gamma$$

which is enough to complete the proof of Theorem 15.7.

15.5   Exercises

15.1 Let $S(\phi)$ be any formula shape for which

$$\vdash_K S(\phi) \rightarrow (\,\Box\phi \rightarrow \phi)$$

holds for all formulas ϕ. Let S be the formal system axiomatized by all the formulas

$$\Box S(\phi) \rightarrow \Box\phi$$

(for arbitrary ϕ).

(a) Show that $S \leq LL$.

(b) Show that S has the rule of disjunction.

(c) Show that the canonical structure $\mathfrak{S}$ of S has an all seeing point.

(d) Does this means that S is not canonical?

15.2 (Grzegorczyk formula) For each formula ϕ let

$$U(\phi) := \Box(\phi \rightarrow \Box\phi) \rightarrow \phi \quad , \quad Grz(\phi) := \Box U(\phi) \rightarrow \phi.$$

(a) For an arbitrary formula ϕ set

$$\theta := \Box\phi \rightarrow \Box^2\phi \quad , \quad \psi := \phi \wedge \theta.$$

Show that the following hold where in (i) the formula ξ is arbitrary.

(i) $\vdash_K (\psi \rightarrow \xi) \rightarrow (\phi \rightarrow \Box\phi \vee \xi)$
(ii) $\vdash_K \Box\psi \rightarrow \Box\phi$
(iii) $\vdash_K (\psi \rightarrow \Box\psi) \rightarrow (\phi \rightarrow \Box\phi)$
(iv) $\vdash_K \Box(\psi \rightarrow \Box\psi) \rightarrow \theta$
(v) $\vdash_K \phi \rightarrow U(\psi).$

(b) Let S be any standard formal system which entails all the formulas $Grz(\phi)$ for arbitrary ϕ. Show that for all formulas ϕ both

$$\vdash_S \Box\phi \rightarrow \phi \quad , \quad \vdash_S \Box\phi \rightarrow \Box^2\phi$$

hold.

15.3 Using the notation of the previous Exercise, let

$$Hrz(\phi) := \Box U(\phi) \rightarrow \Box\phi$$

(for arbitrary ϕ). let

$$KG \quad , \quad KH$$

be the formal systems axiomatized by all the formulas

$$\mathrm{Grz}(\phi) \quad , \quad \mathrm{Hrz}(\phi)$$

respectively. Let **KHT** be the extension of **KH** given by the addition of the axiom T.

(a) Show that **KG** = **KHT**.

(b) Show that **KH** has the rule of disjunction.

(c) Show that **KG** has the rule of disjunction.

(d) Find a formula ϕ such that both

$$\vdash_{\mathsf{KG}} \phi \quad , \quad \neg[\vdash_{\mathsf{KH}} \phi]$$

hold.

Chapter 16

Canonicity without the fmp

16.1 Introduction

In earlier chapters we have isolated two structural properties of a standard formal system S, namely being canonical and having the fmp, both of which ensure that S has the important property of being Kripke-complete. It is, therefore, of interest to ask whether these two structural properties are equivalent. Well, they aren't, in fact they are independent. In the previous two chapters we have seen two systems SLL and LL which have the fmp but are not canonical. In this chapter we will see a system which is canonical but does not have the fmp. Initially this system was constructed solely to distinguish between these two properties but, as we will see, it does have some independent interest.

16.2 The system defined

We work in a monomodal language, so we are interested in transition structures of the form
$$\mathcal{A} = (A, \longrightarrow).$$
Let T be that standard formal system whose axioms are all the formulas of the shapes

(T) $\quad \Box \phi \rightarrow \phi$

(∗) $\quad \Box(\Box^2\phi \rightarrow \sqcup\psi) \rightarrow (\Box\phi \rightarrow \psi)$

where ϕ and ψ are arbitrary. We will see that T is Kripke-complete but does not have the fmp.

The axiom T is, of course, the one that captures the reflexivity of the transition relation. The structural property captured by (∗) will be described in the next section.

By setting $\psi = \Box^2\phi$ in (∗) we obtain the shape

(∗ ∗ ∗) $\quad \Box(\Box^2\phi \rightarrow \Box^3\phi) \rightarrow (\Box\phi \rightarrow \Box^2\phi)$

217

which in some ways is easier to understand. Let R be the system whose axioms are all formulas T and (∗ ∗ ∗). Clearly

$$R \leq T$$

and we will see that R does not have the fmp. However, the other properties of R are more enigmatic.

16.3   The characteristic properties

As with the proof of many results about standard formal systems, the crucial step is to obtain a correspondence property. In the case of T this property is rather strange.

16.1 LEMMA. *A transition structure* $\mathcal{A} = (A, \longrightarrow)$ *models the shape* (∗) *if and only if the property:*

> *For each element* $a \in A$, *there is some* $b \in A$ *with*
>
> - $a \longrightarrow b \longrightarrow a$
> - *for all* $x, y \in A$, $b \longrightarrow y \longrightarrow x \Rightarrow a \longrightarrow x$.

holds.

Proof. Suppose first that $\mathcal{A}$ models the shape (∗), in particular $\mathcal{A}$ models

$$\Box(\Box^2 P \to \Box Q) \to (\Box P \to Q)$$

where P and Q are two fixed and distinct variables. Consider any given element $a \in A$ and let α be any valuation on $\mathcal{A}$ such that

$$x \Vdash P \Leftrightarrow a \longrightarrow x \quad , \quad x \Vdash Q \Leftrightarrow x \neq a$$

hold for all $x \in A$. Then

$$a \Vdash \Box P \quad , \quad a \Vdash \neg Q$$

so that

$$a \Vdash \neg(\Box P \to Q)$$

and hence, using the modelled formula,

$$a \Vdash \Diamond(\Box^2 P \wedge \Diamond \neg Q).$$

This gives some $b \in A$ with $a \longrightarrow b$ and

$$b \Vdash \Box^2 P \quad , \quad b \Vdash \Diamond \neg Q$$

from which the two required listed items follow.

Secondly, suppose that $\mathcal{A}$ has the described structural property, and consider any valuation α on $\mathcal{A}$ and element a of $\mathcal{A}$ with

$$a \Vdash \Box(\Box^2\phi \to \Box\psi) \quad , \quad a \Vdash \Box\phi$$

for given formulas ϕ and ψ. We require, of course, that $a \Vdash \psi$.

The structural property provides an element b of $\mathcal{A}$ satisfying both the listed items. Since $a \longrightarrow b$ we have

$$b \Vdash \Box^2 \to \Box\psi.$$

Also, for each $x, y \in A$,

$$b \longrightarrow y \longrightarrow x \Rightarrow a \longrightarrow x \Rightarrow x \Vdash \phi$$

so that $b \Vdash \Box^2\phi$, and hence $b \Vdash \Box\psi$. Finally, since $b \longrightarrow a$, we have $a \Vdash \psi$, as required. ∎

16.4 Canonicity

With the correspondence result of the previous section, we can attempt the usual proof.

16.2 THEOREM. *The system* T *is canonical, and hence Kripke-complete.*

Proof. Let

$$\mathfrak{T} = (T, \longrightarrow)$$

be the canonical structure associated with T based on the set T of all maximally T-consistent sets of formulas. A routine argument shows that $\mathfrak{T}$ is reflexive, so it remains to show that $\mathfrak{T}$ has the structural property given in Lemma 16.1.

Consider any $s \in T$. We must produce an appropriate $t \in T$. To this end consider the set of formulas

$$\Phi = \{ \Box^2\phi \mid \Box\phi \in s \} \cup \{ \Diamond\theta \mid \theta \in s \}.$$

We show first that Φ is T-consistent.

By way of contradiction suppose that Φ is not T-consistent. Then there are appropriate formulas $\phi_1, \ldots, \phi_m, \theta_1, \ldots, \theta_n$, with

$$\vdash_T \Box^2\phi_1 \wedge \cdots \wedge \Box^2\phi_m \wedge \Diamond\theta_1 \wedge \cdots \wedge \Diamond\theta_n \to \bot.$$

Let

$$\phi = \phi_1 \wedge \cdots \wedge \phi_m \quad , \quad \theta = \theta_1 \wedge \cdots \wedge \theta_n.$$

Then
$$\Box \phi \in s \quad , \quad \theta \in s$$
and
$$\vdash_T \; \Box^2 \phi \; \leftrightarrow \; \Box^2 \phi_1 \wedge \cdots \wedge \Box^2 \phi_m \quad , \quad \vdash_T \; \Diamond \theta \; \rightarrow \; \Diamond \theta_1 \wedge \cdots \wedge \Diamond \theta_n$$
and hence
$$\vdash_T \; \Box^2 \phi \wedge \Diamond \theta \rightarrow \perp.$$
This gives
$$\vdash_T \; \Box^2 \phi \rightarrow \Box \neg \theta$$
so that
$$\vdash_T \; \Box (\Box^2 \phi \rightarrow \Box \neg \theta)$$
and hence the axiom (∗) gives
$$\vdash_T \; \Box \phi \rightarrow \neg \theta$$

which in turn gives the contradictory $\neg \theta \in s$.

This shows that Φ is T-consistent, and hence there is some $t \in T$ with $\Phi \subseteq t$. For any such t and formula ϕ we have, using the axiom T,
$$\Box \phi \in s \; \Rightarrow \; \Box^2 \phi \in \Phi \subseteq t \; \Rightarrow \; \phi \in t$$

so that $s \longrightarrow t$. Also, for each formula θ,
$$\theta \in s \; \Rightarrow \; \Diamond \theta \in \Phi \subseteq t$$

so that $t \longrightarrow s$. Finally, consider any $u, v \in S$ with
$$t \longrightarrow v \longrightarrow u.$$

then, for each formula ϕ,
$$\Box \phi \in s \; \Rightarrow \; \Box^2 \phi \in t \; \Rightarrow \; t \in u$$

so that $s \longrightarrow u$, as required. ∎

16.5   The finite models

In this section we consider the finite models of R. Since any model of T is also a model of R this will give us some information about the finite models of T.

16.3 LEMMA. *Each finite model of* R *is transitive.*

Proof. Let $\mathcal{A}$ be any model of R and consider the induced modal operation $\square$ on $\mathcal{P}A$. Since the transition relation of $\mathcal{A}$ is reflexive, the operation $\square$ is deflationary, and hence $\mathcal{A}$ is transitive precisely when $\square$ is idempotent. Consider any $X \subseteq A$. Since $\square$ is deflationary we have a descending chain

$$X \supseteq \square X \supseteq \square^2 X \supseteq \cdots \supseteq \square^r X \supseteq \cdots \qquad (r < \omega).$$

Suppose that

$$\square^3 X = \square X.$$

Then

$$\square(\square^2 X \supset \square^3 X) = A$$

so that, using axiom $(***)$, we have

$$(\square X \supset \square^2 X) = A$$

and hence

$$\square^2 X = \square X.$$

An easy induction now shows that

$$\square^{r+3} X = \square^{r+1} X \;\Rightarrow\; \square^2 X = \square X$$

for each $r < \omega$.

Finally, if $\mathcal{A}$ is finite, then the descending chain given above must stabilize, i.e. there is some $r < \omega$ with $\square^{r+3} X = \square^{r+1} X$. Thus $\square^2 X = \square X$, which is enough to show transitivity. ∎

Let $\mathbb{M}$ be the class of all finite models of T. Thus each member of $\mathbb{M}$ is transitive and so models

$$\square\phi \to \square^2 \phi$$

for each formula ϕ. Thus if T has the fmp, then

$$\vdash_{\mathsf{T}} \square\phi \to \square^2 \phi$$

and so every model of T is transitive. In the next section we show this is not so, and hence prove the following.

16.4 THEOREM. *The system* T *is canonical (and hence Kripke-complete) but does not have the fmp.*

16.6   A particular model

Consider the structure

$$\mathcal{N} = (\mathsf{N}, \longrightarrow)$$

where the relation $\longrightarrow$ is defined by

$$x \longrightarrow y \iff x \leq y + 1$$

(for each $x, y \in \mathsf{N}$). Clearly $\mathcal{N}$ is reflexive.
    Consider any $a \in \mathsf{N}$ and let $b = a + 1$. Then

$$b \longrightarrow a \longrightarrow a.$$

Also, for each $x, y \in \mathsf{N}$, we have

$$b \longrightarrow y \longrightarrow x \Rightarrow b \leq y + 1 \leq x + 2 \Rightarrow a \leq y \leq x + 1 \Rightarrow a \longrightarrow x.$$

This shows that $\mathcal{N}$ has the structural property of Lemma 16.1 and hence is a model of T. However,

$$4 \longrightarrow 3 \longrightarrow 2 \quad \text{but} \quad \neg[4 \longrightarrow 2]$$

so that $\mathcal{N}$ is not transitive, and hence T has a non-transitive model.

16.7   Exercises

16.1 Show that $\mathsf{T} \leq \mathsf{S4}$.

16.2 For $i, j, k, l \in \mathsf{N}$ with

$$2 \leq j < k \quad , \quad l + 1 \leq i + j$$

let $\mathsf{R}(i, j, k, l)$ and $\mathsf{S}(i, j, k, l)$ be the formal systems axiomatized by T together with, respectively, all formulas of the shape

$$\Box^i(\Box^j\phi \to \Box^k\phi) \to \Box^l(\Box\phi \to \Box^2\phi)$$

or

$$\Box^i(\Box^j\phi \to \Box^k\psi) \to \Box^l(\Box\phi \to \Box^2\psi)$$

for arbitrary ϕ, ψ. (Thus the system R of this chapter is $\mathsf{R}(1, 2, 3, 0)$.)

(a) Show that $\mathsf{R}(i, j, k, l) \leq \mathsf{S}(i, j, k, l)$.

(b) Show that if

$$i \geq i' , \ j \leq j' , \ k \geq k' , \ l \leq l'$$

then both

$$\mathsf{R}(i', j', k', l') \leq \mathsf{R}(i, j, k, l) \quad \text{and} \quad \mathsf{S}(i', j', k', l') \leq \mathsf{S}(i, j, k, l)$$

hold.

(c) Show that each finite model of $R(i,j,k,l)$ is transitive.

(d) Recalling the correspondence result of Exercise 5.3 of Chapter 5, show that $S(i,j,k,l)$ is canonical.

(e) Show that the structure $\mathcal{N}$ of Section 16.6 models $S(i,j,k,l)$.

(f) Show that neither $R(i,j,k,l)$ nor $S(i,j,k,l)$ has the fmp.

Chapter 17

Transition structures aren't enough

17.1 Introduction

The central theme of this book is the notion of a labelled transition structure and the use of modal languages to analyse such structures. We have seen many examples of formal systems S and, in Chapter 9, we have proved a quite general completeness result for such systems in terms of the semantics supported by labelled transition structures. We have also developed methods (mainly canonicity and the fmp) for obtaining the stronger property of Kripke-completeness. In fact we have yet to see an example of a system which is not Kripke-complete. In this chapter I will give such a system. This example will also indicate some of the discrepancies between labelled transition structures and modal systems.

I will describe two standard formal systems

$$\mathsf{KY} \quad , \quad \mathsf{KZ}$$

with the following properties.

- KZ is canonical and has the fmp.

- $\mathsf{KY} \leq \mathsf{KZ}$.

- KY and KZ have the same class of unadorned models.

- KY and KZ have distinct classes of valued models.

The first property is achieved, using Exercises 10.1 of Chapter 10 and 13.1 of Chapter 13, since KZ is axiomatized by a single sentence. The first three of these properties ensure that, for each formula ϕ, the following scheme of ten

225

implications hold.

$$\vdash_{\mathsf{KY}} \phi \Longleftrightarrow \models^v_{\mathsf{KY}} \phi \overset{*}{\Longrightarrow} \models^u_{\mathsf{KY}} \phi$$

$$*\Big\Downarrow \qquad\qquad\qquad \Big\Uparrow\Big\Downarrow$$

$$\vdash_{\mathsf{KZ}} \phi \Longleftrightarrow \models^v_{\mathsf{KZ}} \phi \Longleftrightarrow \models^u_{\mathsf{KZ}} \phi$$

The fourth property ensures that neither of the implications $(*)$ is reversible, in particular, the system KY is not Kripke-complete.

17.2   The system KZ

We work in the monomodal language and we use the informal terminology

$$a \text{ can see } b \quad \Leftrightarrow \quad a \longrightarrow b$$
$$a \text{ is blind} \quad \Leftrightarrow \quad a \text{ can see no elements}$$

for elements a and b of the considered structure. In particular

$$a \Vdash \square\bot \quad \Leftrightarrow \quad a \text{ is blind}$$
$$a \Vdash \lozenge\square\bot \quad \Leftrightarrow \quad a \text{ can see a blind element}$$

Consider the sentence

$$Z := \square\lozenge\top \to \square\bot$$

and let KZ be the system whose sole proper axiom is Z.

As remarked above, we know that KZ is canonical and has the fmp. Note also, since KZ is axiomatized by a sentence, a valued structure $(\mathcal{A}, \alpha)$ models KZ precisely when $\mathcal{A}$ models KZ (for the valuation plays no part in ther modelling process). We can easily characterize the unadorned models of KZ. The above discussion gives the following

17.1   LEMMA. *Each modal structure $\mathcal{A}$ is a model of KZ precisely when each non blind element can see a blind element.*

For each formula θ let

$$Z(\theta) := \square\lozenge\top \to \square\theta$$

so that Z is just $Z(\bot)$. Note also that

$$\vdash_{\mathsf{K}} \bot \to \theta$$

so that (EN) gives

$$\vdash_{\mathsf{K}} \square\bot \to \square\theta$$

and hence, using the axiom, we have

$$\vdash_{\mathsf{KZ}} \square\lozenge\top \to \square\theta \quad \text{i.e.} \quad \vdash_{\mathsf{KZ}} Z(\theta).$$

This enables us to construct many systems weaker than KZ by taking axioms of the form $Z(\theta)$ for various shapes of θ. The system KY is one of these.

17.3   The system KY

For each formula ϕ, we have

$$\mathrm{T}(\phi) \quad := \quad \Box\phi \to \phi.$$

We use this to construct the formulas

$$\mathrm{X}(\phi) := \Box\mathrm{T}(\phi) \to \phi \quad \text{and} \quad \mathrm{Y}(\phi) := \Box\Diamond\mathrm{T} \to \Box\mathrm{X}(\phi).$$

In the above notation $\mathrm{Y}(\phi)$ is just $\mathrm{Z}(\mathrm{X}(\phi))$. Let KY be the system whose set of proper axioms are all the formulas $\mathrm{Y}(\phi)$ (for arbitrary ϕ). As a particular case of the observation at the end of the last section we have

$$\vdash_{\mathsf{KZ}} Y(\phi)$$

which demonstrates the following.

17.2   PROPOSITION. KY $\leq$ KZ.

The crucial result is to connect the class of models of KY and KZ. This is done by the next result (which is a correspondence result for the shape Y).

17.3   THEOREM. *For each structure $\mathcal{A}$ the conditions*

(i) $\mathcal{A} \Vdash^u Z$.
(ii)   For all formulas ϕ, $\mathcal{A} \Vdash Y(\phi)$.
(iii)   For some variable P, $\mathcal{A} \Vdash Y(P)$.

are equivalent.

Proof. (i) $\Rightarrow$ (ii). This follows by soundness and Proposition 17.2.
(ii) $\Rightarrow$ (iii). This is trivial.
(iii) $\Rightarrow$ (i). Consider any configuration

$$a \longrightarrow b$$

together with a valuation on $\mathcal{A}$ such that for each $x \in A$

$$x \Vdash P \Leftrightarrow x \neq b$$

holds (for the given variable P). Then for each x with $b \longrightarrow x$ we have either

$$x \neq b \quad \text{or} \quad x = b.$$

In the first case we have $x \Vdash P$ and hence $x \Vdash \mathrm{T}(P)$. In the second case have

$$x = b \longrightarrow x = b$$

so that $x \Vdash \Diamond \neg P$, and again $x \Vdash T(P)$. In both cases we have $x \Vdash T(P)$ which shows that

$$b \Vdash \Box T(P).$$

Now, assuming (iii) we have

$$a \Vdash Y(P)$$

and hence

$$a \Vdash \Box \Diamond \top \ \Rightarrow \ a \Vdash \Box X(P)$$
$$\Rightarrow \ b \Vdash X(P)$$
$$\Rightarrow \ b \Vdash P \qquad \Rightarrow \ b \neq b$$

so that $a \Vdash \neg \Box \Diamond \top$.

This shows that for each $a \in A$

$$a \Vdash \Diamond \top \rightarrow \neg \Box \Diamond \top$$

or, in contrapositive,

$$a \Vdash \Box \Diamond \top \rightarrow \Box \top$$

as required. ∎

This result gives the third of four properties listed in Section 17.1. It remains to verify the fourth.

17.4   A particular structure

In this section we construct a particular modal structure $\mathcal{A}$ and analyse some of its properties. This structure will provide an appropriate example to show that the two systems KY and KZ are distinct.

To begin the construction let

$$A = \mathsf{N} \cup \{\omega, \infty\}$$

i.e. let A be the set of natural numbers N together with two extra elements which we choose to call 'ω' and '∞'. We enrich A with a simple transition relation $\longrightarrow$ to form the structure $\mathcal{A}$. Thus, for $m, n \in \mathsf{N}$ we let

$$n \longrightarrow m \ \Leftrightarrow \ m < n$$

and, for each $m \in \mathsf{N}$, we let

$$\omega \longrightarrow m$$

hold together with

$$\infty \longrightarrow \omega.$$

No other instances of $\longrightarrow$ hold. In particular, neither ∞ nor ω is blind (since ∞ can see ω, and ω can see 0), and ω is the only world seen by ∞. This shows that

- $\mathcal{A}$ is not a model of KZ.

The remainder of this section is concerned with showing the following.

- For many valuations α, the structure $(\mathcal{A}, \alpha)$ is a model of KY.

In particular, there is at least one such valuation α. Then $(\mathcal{A}, \alpha)$ does not model KZ but does model KY, and hence KY and KZ have distinct classes of valued models.

We use a certain class $\mathcal{X}$ of subsets of $\mathcal{A}$ (i.e. $\mathcal{X} \subseteq \mathcal{P}A$). Thus let $\mathcal{X}$ be the set of all subsets U of A such that either

$$(+) \qquad U \text{ is finite with } \omega \notin U$$

or

$$(-) \qquad U' \text{ is finite with } \omega \in U$$

(where here U' is the complement $A - U$).

Observe that

$$\emptyset \quad , \quad A \quad , \quad \{\infty\} \quad , \quad \mathsf{N} \cup \{\omega\}$$

are all in $\mathcal{X}$ (for the first and third satisfy $(+)$, and the second and fourth satisfy $(-)$). Note also that neither $\{\omega\}$ nor $\{\omega, \infty\}$ are in $\mathcal{X}$. Next you should check that, for each subset U of A,

$$U \text{ satisfies one of } (+)\,(-) \quad \Leftrightarrow \quad U' \text{ satisfies the other}$$

so that $\mathcal{X}$ is closed under complementation. Finally, a few simple computations show that $\mathcal{X}$ is closed under $\cup$ and $\cap$. For instance, suppose that U has $(+)$ and V has $(-)$. Then

$$\omega \in V \subseteq U \cup V \quad \text{and} \quad (U \cup V)' = U' \cap V' \subseteq V'$$

so that $U \cup V$ has $(-)$.

This shows that $\mathcal{X}$ is closed under the boolean operations $\cup, \cap, (\cdot)'$ (which, of course are used to interpret the boolean connectives of the language). We add to this a similar, but stronger, result concerning the induced box operation $\square$ on $\mathcal{P}A$.

17.4 LEMMA. *For each subset U of A we have $\square U \in \mathcal{X}$ (and hence $\mathcal{X}$ is closed under the operation $\square$).*

Proof. Consider any $U \subseteq A$ and recall that for all $a \in A$

$$a \in \square U \quad \Rightarrow \quad (\forall x \prec a)[x \in U].$$

To compute $\square U$ we consider two cases.

Firstly, suppose that $\mathsf{N} \subseteq U$. Then U is one of

$$\mathsf{N} \quad , \quad \mathsf{N} \cup \{\omega\} \quad , \quad \mathsf{N} \cup \{\infty\} \quad , \quad A$$

and a simple calculation shows that

$$\Box N = \Box(N \cup \{\infty\}) = N \cup \{\omega\} \quad \text{and} \quad \Box(N \cup \{\omega\} = \Box A = A.$$

Secondly, suppose that $N \nsubseteq U$, so there is a least $n \in N-U$. Then, for each $m \in N$ we have

$$m \in \Box U \;\Rightarrow\; (\forall x < m)[x \in U] \;\Rightarrow\; n \nleq m$$

i.e.

$$m \in \Box U \;\Rightarrow\; m \leq n.$$

Also, we easily check that

$$0, 1, 2, \cdots, n \in \Box U$$

and hence, with a little more computation, we find that

$$\Box U = \begin{cases} \{0, \cdots, n, \infty\} & \text{if } \omega \in U \\[2mm] \{0, \cdots, n\} & \text{if } \omega \notin U. \end{cases}$$

In both cases $\Box U$ has $(+)$, and hence $\Box U \in \mathcal{X}$. ∎

You should make sure that you understand the computations of this proof, for we are going to use several more similar computations.

First a simple observation

17.5 PROPOSITION. *Let α be any valuation on $\mathcal{A}$ such that*

$$\alpha(P) \in \mathcal{X}$$

for all $P \in Var$. Then

$$[\phi]_\alpha \in \mathcal{X}$$

for all formulas ϕ.

Proof. This follows by a routine induction on the complexity of ϕ using Lemma 17.4 and the fact that $\mathcal{X}$ is closed under the boolean operations. ∎

Let us now do some computations of

$$[\phi]_\alpha$$

for particular cases of ϕ and α.

Note first that, from the computation in the proof of Lemma 17.4, we have

$$\Box \emptyset = \{0\}$$

and hence (for any valuation α)

$$
\begin{aligned}
[\![\Box\Diamond\mathsf{T}]\!] &= \Box\Diamond[\![\mathsf{T}]\!] \\
&= \Box\Diamond A \\
&= \Box\neg\Box\emptyset \\
&= \Box\{1,\cdots,\omega,\infty\} = \{0,\infty\}.
\end{aligned}
$$

Next consider any formula ϕ and valuation with

$$U = [\![\phi]\!] \in \mathcal{X}$$

(where we have omitted the distinguishing valuation). We will compute

$$[\![\mathsf{T}(\phi)]\!] \quad , \quad [\![\mathsf{X}(\phi)]\!] \quad , \quad [\![\,\Box\mathsf{X}(\phi)]\!]$$

in terms of U. To do this it is convenient to isolate several cases.

In the first case we suppose

(i) $\mathsf{N} \subseteq U$ (so that, since $U \in \mathcal{X}$, also $\omega \in U$).

Secondly, when (i) isn't the case there is some least $n \in \mathsf{N}-U$ and with this n we have three further cases.

(ii) $n \notin U$, $\omega \in U$, $\infty \in U$
(iii) $n \notin U$, $\omega \in U$, $\infty \notin U$
(iv) $n \notin U$, $\omega \notin U$

The calculations in the four cases are similar but slightly different as follows.

First, by the previous calculation, we have

$$
\Box U = \begin{cases}
A & \text{in case (i)} \\
\{0,\cdots,n,\infty\} & \text{in cases (ii,iii)} \\
\{0,\cdots,n\} & \text{in case (iv)}
\end{cases}
$$

so that, since

$$[\![\mathsf{T}(\phi)]\!] = (\Box U)' \cup U$$

we have

$$
[\![\mathsf{T}(\phi)]\!] = \begin{cases}
U & = U & \text{in case (i)} \\
U \cup \{n+1,\cdots,\omega\} & = \{n\}' & \text{in case (ii)} \\
U \cup \{n+1,\cdots,\omega\} & = \{n,\infty\}' & \text{in case (iii)} \\
U \cup \{n+1,\cdots,\omega,\infty\} & = \{n\}' & \text{in case (iv)}
\end{cases}
$$

and hence

$$
[\![\,\Box\mathsf{T}(\phi)]\!] = \Box[\![\mathsf{T}(\phi)]\!] = \begin{cases}
A & \text{in case (i)} \\
\{0,\cdots,n,\infty\} & \text{in cases (ii, iii, iv).}
\end{cases}
$$

Continuing in the same way, since

$$[\![\mathsf{X}(\phi)]\!] = ([\![\,\Box\mathsf{T}(\phi)]\!])' \cup [\![\phi]\!]$$

we find that

$$[\![X(\phi)]\!] = \begin{cases} U & \text{in case (i)} \\ U \cup \{n+1, \cdots, \omega\} & \text{in cases(ii,iii,iv)} \end{cases}$$

and hence

$$[\![\,\Box X(\phi)]\!] = \begin{cases} A & \text{in case (i)} \\ \{0, \cdots, n, \infty\} & \text{in cases(ii,iii,iv).} \end{cases}$$

These computations provide much of the proof of the following result.

17.6 PROPOSITION. *For each valuation* α, *if* $\alpha(P) \in \mathcal{X}$ *for all variables* P, *then* $(\mathcal{A}, \alpha)$ *models the system* **KY**.

Proof. Consider any formula ϕ. By Proposition 17.5 we have

$$U = [\![\phi]\!]_\alpha \in \mathcal{X}$$

so that, from above,

$$\{0, \infty\} \subseteq [\![\,\Box X(\phi)]\!]$$

(where again we have dropped the α). Thus, remembering the computed value of $[\![\,\Box \Diamond \top]\!]$, we have

$$[\![Y(\phi)]\!] = A$$

as required. ∎

17.5   Exercises

17.1 Consider the structure $\mathcal{A}$ of Section 17.4.

(a) Compute the following.

   (i) $[\![\,\Box^{r+1} \bot]\!]$ (ii) $[\![\,\Box \Diamond \bot]\!]$ (iii) $[\![\,\Box \Diamond \Box \top]\!]$
   (iv) $[\![\,\Box \Diamond \Box \bot]\!]$ (v) $[\![Z(\theta)]\!]$ (vi) $[\![4(\theta)]\!]$

   where, in (v,vi), θ is an arbitrary formula.

(b) Find

   (i) all $U \subseteq A$ such that $\Box U = U$,
   (ii) some $U \subseteq A$ such that $\Box U = \neg U$,
   (iii) some $U \subseteq A$ such that all the sets $\Box^r U$, for $r < \omega$, are distinct.

(c) Show that $\Box$ is not continuous in the sense that there is an ascending sequence $\mathcal{U}$

$$U_0 \subseteq U_1 \subseteq \cdots \subseteq U_r \subseteq \cdots \quad r < \omega$$

of sets such that

$$\Box \bigcup \mathcal{U} \neq \bigcup \Box \mathcal{U}$$

where $\Box \mathcal{U} = \{ \Box U_r \mid r < \omega \}$.

17.2 We have seen that there are infinite valued structures $(\mathcal{A}, \alpha)$ which model KY but not KZ. Are there finite valued structures which can do the same job?

17.3 Here is another example of a formal system which is not Kripke-complete; this time a bimodal one.

Let S be the extension of the basic temporal system TEMP (of Chapter 8, Section 8.4) formed by the addition of the two axiom shapes

$$L_-(\phi) \quad , \quad M_+(\phi)$$

for arbitrary formulas ϕ (where L_- is Löb's axiom for $[-]$, and M_+ is McKinsey's axiom for $[+]$).

(a) Using Theorem 5.7 and Exercise 5.8 of Chapter 5, show that no temporal structure models S.

(b) Consider the temporal structure $\mathcal{N} = (\mathsf{N}, <)$

   (i) Show that $\mathcal{N}$ models L_-.

   (ii) Show that for each $X \subseteq \mathsf{N}$

$$X \text{ finite} \Rightarrow [+]\langle + \rangle X = \emptyset \quad , \quad X \text{ infinite} \Rightarrow [+]\langle + \rangle X = \mathsf{N}.$$

   (iii) Show that if ν is a valuation on $\mathcal{N}$ such that for each variable P either $\nu(P)$ is finite or $\nu(P)$ is cofinite then $(\mathcal{N}, \nu)$ models S.

(c) Show that S is not Kripke-complete.

Part VI

Two appendices

Appendix A

The what, why, where,... of modal logic

A.1 Introduction

This chapter is one of the first parts of this book that you should read. You might think this is a little strange since it is also one of the last things in the book. However, all of the material in this chapter should appear somewhere, and I believe that it is better if it is all in one place rather than scattered throughout the book. This position is as good as any.

A.2 Beginning

Put yourself in the position of a complete beginner to modal logic; someone who already knows the basics of propositional and predicate logic and who now wants to learn something of modal logic. (You may actually be such a person.) There are many reasons why you may want to do this, from mere curiosity to an eventual use in a particular application. What should you do to acquire this knowledge?

One thing you could do is attend a course on modal logic, but let us assume that this is not an option. The other thing to do is read various text books on the subject. Which ones? There are, in fact, only a few possibilities.

The first possibility is [29] by Hughes and Cresswell, first published in 1968. This, at the time it was written, was the most comprehensive and accessible introduction to the subject. It contains descriptions of many of the systems that were, and to some extent still are, of interest. It discusses the proof theoretic properties, completeness, and decidability of these systems, and various comparisons between them. (It also contains some material on predicate modal logic.) This book will give you a good impression of what modal logic used to be.

However, the book was written before the use of Kripke structures was fully understood and consequently is severely limited in technique and, in terms of content, seems a little ad hoc.

The second book by Hughes and Cresswell, [30], published some 16 years later, benefits from the influence of Kripke's insight. The difference in style between the two presentations is immediately apparent. The aim of the book is more or less a standard introduction to the subject without unnecessarily repeating material from [29]. Thus, along with the basics it contains material on Kripke semantics, completeness, correspondence results, and other structural properties. (It again also contains some predicate modal logic.) This book, perhaps augmented by material from [29], can be used as an entry into modal logic.

The book [17] (which chronologically separates [29] and [30]) has probably been the standard introduction to the subject since it was published (in 1980). It contains much of what is needed; a survey of the various systems of interest, completeness properties, correspondence results, decidability, etc. It can be used either as an alternative to [30] or, better still, combined with [30] to produce a balanced account. Like much of the literature on modal logic, these books concentrate too much on syntactic considerations, enumerating the properties of many different systems and the interaction between them. Semantical aspects are always a long way behind.

Two other books you may like to consult are [38] and [42]. Unfortunately, these are sometimes rather difficult to get hold of.

All of these books have, from the present perspective, one major flaw: they are all concerned with monomodal logic. This gives them what is increasingly being seen as a rather peculiar slant. It also cuts out a lot of material which ought to be part of the basics of modal logic. The subject ought to be seen as an analysis of the interaction between various kinds of modalities; in short the underlying language ought to be polymodal.

A more balanced view of the subject is given in the extended survey [27]. This little book is one of the best places to start to learn the subject. It quickly and succinctly describes the pertinent techniques, doesn't linger too long over exotic proof theoretic considerations, and gives Kripke structures the central place they deserve. You will, however, need to augment this book with other material.

The book you are now reading is an attempt to meet these various objections to the presently available accounts of modal logic. In doing this it starts from a fundamentally different point of view. The central objects of study are not the syntactic systems of modal logic; these are merely the tools of the trade. These tools are used to analyse and describe the basic objects of interest which are **Labelled Transition Structures** (sets furnished with a family of binary relations).

Particular kinds of LTSs occur in many different areas. Kripke structures are themselves (unlabelled) TSs and so many 'possible world' situations provide examples of LTSs. Any mechanism that has different states and transitions between these states is an LTS. Such mechanisms occur in program semantics (in the form of abstract versions of the machines on which the programs are executed, and the various rewriting systems used to compile and describe the operational behaviour of these programs). LTSs are the supporting structures on which many of the accounts of distributed communicating systems are based. Aspects of LTSs can be found in the analysis of non-well-founded sets and parts of situation theory. More generally, since transition structures are nothing more than collections of binary relations, it is not hard to find examples of them all over the place.

So we decide to make LTSs our central notion. What material should a first course about these structures contain?

We could go for a purely algebraic account. We could describe such properties as the appropriate notions of morphism between these structures (the standard morphisms, p-morphism, simulations and bisimulations); the various constructions on LTSs (generated substructures, ultrafilter extensions, etc); the related notion of an enriched boolean algebra (modal algebra); and perhaps culminating in a version of the Stone representation theorem for these structures.

Such a course would make no mention at all of modal logic (just as a course on boolean algebras, or more generally, lattices, need not mention propositional logic). All of these topics have to be learned at some time but not necessarily at the beginning. We are concerned here with another important tool – the use of modal languages. (Once you have mastered these basic techniques you should look at the book [10]. There the interplay between the algebraic and logical aspects are developed in depth.)

As with other areas of mathematics, many of the interesting properties of LTSs are describable in a second order fashion. In some mysterious way modal languages capture many of these properties in an apparently zero order fashion. This makes modal languages a powerful tool for the analysis and codification of LTSs. The central objective of modal logic is to develop and refine this tool. Usually this is done with a particular application in mind but there is enough common material to justify a unified account of the basics of the subject.

A.3 About this book

You want to learn some modal logic and after a bit of dithering you have decided to be sensible and use this book. What can you expect from the book, and what will the book expect of you?

Let me first set down the perimeters inside which the book operates.

- The book is an *introduction* to modal logic. It assumes no prior knowledge of the subject, but does assume some familiarity with propositional logic (and a little bit of predicate logic). It will also help if you already have a bit of mathematical sophistication, if not you should be prepared to acquire some as you read.

- Because the book is an introduction, it can not be a comprehensive account of the subject, even of the basic parts of the subject. In order to keep it down to a reasonable size I have had to be selective. The main criteria used to make these selections is that I believe that semantical considerations should come first.

- The book is concerned exclusively with propositional modal logic. Predicate enrichments are not dealt with. Furthermore, the base logic is classical (i.e 2-valued). Modal enrichments of constructive logic are not considered. This excludes a lot of material but does, I believe, stay within a natural boundary.

- The central concept of the book is that of a Labelled Transition Structure and modal languages are regarded as a tool for analysing these structures. This means that the main emphasis is on semantical aspects. Proof theoretical methods are barely touched on. This, I believe is the correct perspective for an introductory text. Proof theoretic manipulations can become meaningless gibberish without some supporting semantics to aid, at least, the soundness of the methods. Transition structures provide almost all of these supporting semantics.

I have organized the book so that it could be read linearly, starting at the beginning and working through to the end. However, as with almost all mathematical texts, I wouldn't read it like that. You should read the book in such a way that you get to the heart of the material as quickly as possible, missing out details on the way, but frequently going back to fill in these details as necessary.

A possible route through the material is:

- Parts of Chapter 2;

- Parts of Chapter 3;

- Parts of Chapter 4;

followed by either

- Parts of Chapter 5;

or

- Parts of Chapter 8.

You are then in a position to go through the proof of a reasonably non-trivial result, say either

- The confluence result, Theorem 5.4

or

- The completeness result, Theorem 9.1.

You will find that to do this you will need to fill in various details that you have skipped, so you must go back and do this. In this way you will eventually build up an understanding of the subject.

An important part of this process is doing exercises. One of the first things you should read in any chapter is the selection of exercises, and you should start doing these exercises as quickly as possible. Also you don't necessarily have to do the exercises in the order listed, if there is some exercise you can't do, then leave it for later and go on to another one.

A fairly extensive set of solutions is given in the Appendix B. Use these initially to get hints to the solutions and only read the full solution after you have sweated over the exercise for a while. It is not enough merely to understand the solution to a problem, you have to learn to produce these solutions yourself, and the only way to do this is by hard work.

A.4 What next?

Suppose now that you have become familiar with the basics of the subject (say parts I, II, III, and IV of this book) and have dipped into the various other texts mentioned so far. You are now ready to stretch yourself a bit further. What should you do next?

Reading the two survey articles [15] and [11] should be high on your list.

The first of these articles [15] describes some of the advanced techniques of monomodal logic (perceived as something to be studied for its own sake). Unfortunately the article does at time read like an aimless ramble through the modal forest rather than a directed tour lead by a guide who knows where he his going. (For a possible explanation of this see the first footnote of the survey.)

The second survey article [11] is concerned with the expressive power of modal languages (again almost exclusively monomodal). Its approach is more in line with this book, i.e. the modal language is there to analyse properties of Kripke structures, not the other way round. If you like this article then you should also spend quite a bit of time on the book [10]. This will help you develop a deeper understanding of modal logic, especially the algebraic aspects. It is written in a monomodal framework, but you should have no difficulty extending much of the book to a polymodal setting.

There are several relevant survey articles in the handbooks [1], [24], [25], and [35], some of which I have already mentioned. The articles [11, 15, 16, 28, 43, 49] of [24] all have something to offer, and the article [18] of [35] is relevant. You will find the articles [6, 48] of [1] of interest with the article by Stirling [48] particularly useful as this will point the way towards many different facets of the subject. In [25] the articles [13, 22, 33] have some promise.

One topic has not been addressed in this book, in fact it has been deliberately avoided.

If modal languages are to be of use then we will inevitably be led to certain particular formal systems. (Which particular systems will depend on the area of application in mind.) At times we will need to give formal proofs within these systems. To be able to do this efficiently a deeper understanding of the workings of these systems need to be developed. If you look at this problem you will quickly realize that the Hilbert style proof systems presented in this book are notoriously difficult to use if actual proofs need to be presented (they were not developed for that purpose). A different, more amenable, style of proof system is required. This problem also occurs with modal-free propositional logic, but it is not so important there. As with modal-free languages there are several possible approaches to efficient modal systems; natural deduction system, sequent systems, tableau systems, etc, each of which can come in many different variants. The arguments for and against these various styles are more concerned with implementation and automation problems rather than the styles themselves, so are better discussed in this setting. If you are interested in these matters you could start with the book [21] and then move on to [50].

A.5   Some uses of modal logic

The original use as a method of analysing the properties of the informal natural language modalities 'is necessary', 'is obligatory', 'is known', ... is still valid. Furthermore, in a polymodal setting the interaction between these constructs can be attacked. In this respect you may like to read [49].

Tense logic as originally conceived is an analysis of tenses in natural languages. An account of this perspective is given in [16], and again you may like to look at [49]. The two books [23] and [9] are specifically aimed at this use of modal logic.

Beginning with Kamp [31] the basic language of tense logic has been enriched, by adding operators such as 'until' ,'since', ..., to produce much more expressive languages and powerful systems. This has spawned the subject of temporal logic which has many different connections with computational problems. A comprehensive survey of temporal logic starting from its origins as tense logic is given in [12]. This survey also contains material on the analysis of temporal properties based on 'intervals' or 'events' as opposed to 'points'. This is something not mentioned in this book.

Temporal logic can also been used to illuminate certain aspects of non-temporal modal logic. To motivate this consider any monomodal structure

$$\mathcal{A} = (A, \longrightarrow)$$

which we assume is serial. A *path* through $\mathcal{A}$ is an infinite sequence of transitions

$$\mathsf{a} := a(0) \longrightarrow a(1) \longrightarrow a(2) \longrightarrow \cdots \longrightarrow a(r) \longrightarrow \cdots$$

$(r < \omega)$. Let A^+ be the set of all pairs (a, r) where a is a path and $r < \omega$. We convert A^+ into a transition structure using the relation

$$(\mathsf{a}, r) \longrightarrow (\mathsf{b}, s+1) \Leftrightarrow a(r) = b(s).$$

Also, given a valuation α on $\mathcal{A}$ we define the valuation α^+ on $\mathcal{A}$ by

$$(\mathsf{a}, r) \in \alpha^+(P) \Leftrightarrow a(r) \in \alpha(P).$$

You may now check that for each $a \in A$ and formula ϕ, the following are equivalent.

- $a \Vdash \phi$

- For all $(\mathsf{a}, r) \in A^+$, if $a(r) = a$ then $(\mathsf{a}, r) \Vdash \phi$.

- There is some path a with $a(0) = a$ and $(\mathsf{a}, 0) \Vdash \phi$.

What is the point of this? Suppose we introduce two new transition relations

$$\xrightarrow{\times} \ , \ \xrightarrow{\circ}$$

on A^+ by

$$(\mathsf{a}, r) \xrightarrow{\times} (\mathsf{b}, s) \Leftrightarrow a(r) = b(s)$$

and

$$(\mathsf{a}, r) \xrightarrow{\circ} (\mathsf{b}, s) \Leftrightarrow \mathsf{a} = \mathsf{b} \text{ and } s = r+1.$$

Then the transition

$$(\mathsf{a}, r) \longrightarrow (\mathsf{b}, s+1)$$

can be decomposed as the composite

$$(\mathsf{a}, r) \xrightarrow{\times} (\mathsf{b}, s) \xrightarrow{\circ} (\mathsf{b}, s+1).$$

Also, introducing the corresponding box operators $[\times]$, $[\circ]$ we find that

$$\Box \phi \leftrightarrow [\times][\circ]\phi \quad \text{and} \quad [\circ]\phi \leftrightarrow \neg[\circ]\neg\phi$$

hold in $\mathcal{A}^+$. Thus we have achieved a decomposition of $\Box$ into two more primitive operators.

A similar construction can be carried out for polymodal systems and languages. This takes us into the realms of linear and branching time temporal logics, a topic which has found uses in various aspects of specification and verification techniques. The article [19] is a full survey of this topic including such aspects as the comparative expressiveness of the various languages used and their decidability properties. There is also a full discussion of the areas of application of these techniques in computing science.

You will find [47] a useful little paper. There you will see some applications of bisimulations to the branching time structure $\mathcal{A}^+$, and a complete axiomatization of the logic of $[\times]$. This completeness proof modifies and extends the standard techniques used in Chapter 9. The survey article [18] deals with these aspects of temporal logic, and the book [40] describes some of its uses.

In several uses of modal logic the set of labels itself carries an algebraic structure which has to be reflected in the behaviour of the modal operators. One example of this is dynamic logic. A brief introduction to this is given in 11.1 of Chapter 11. A full survey can be found in [28].

Another example is the analysis of distributed communicating systems. Much of this analysis is based explicitly or implicitly on the notion of a transition structure. Here the labels (usually called actions in this context) have a rather intricate algebraic structure. The paper [46] is a nice account of how transition structures and various associated modal logics are employed in the study of CCS (one of the calculus used to analyse communicating systems).

Many of the enrichments of the basic modal language can be subsumed under the use of *fixed point operators*. A hint of this can be seen in Exercise 14.4 of Chapter 14. As in that exercise, the semantics of fixed point operators can not be dealt with successfully using only transition structures: it is necessary to use modal algebras (because of their lattice theoretic completeness properties). This is one reason why fixed point constructions have not been discussed in this book. An account of these fixed point enrichments is given in [4].

There is a direct connection between modal logic and automata theory. Given a set I of labels, let I^* be the set of all words on I, i.e. the set of all finite strings

$$\alpha = i_0 i_1 \cdots i_n$$

of labels $i_0, i_1, \ldots, i_n \in I$. Each transition structure of signature I can be used to define certain subsets of I^*. Thus given the structure $\mathcal{A}$ and an element a we say that a *accepts* the word α if there is a path

$$a = a_0 \xrightarrow{i_0} a_1 \xrightarrow{i_1} \cdots \longrightarrow a_n \xrightarrow{i_n} a_{n+1}$$

through $\mathcal{A}$. Let $Acp(a)$ be the set of all words accepted by a. Thus $Acp(a)$ is the regular language defined by a. Notice that for each word α we have

$$\alpha \in Acp(a) \iff a \Vdash \langle \alpha \rangle \top.$$

Thus we see the connection.

Similar ideas can be used to characterize various testing equivalences on process algebras (which are nothing more than transition structures). For a survey of these ideas see [2] (where you will also find connections between modal logic and things you haven't dreamt of).

A general survey of the various applications of modal logic (more specifically, temporal logic) to computing science is given in the book [26].

There are several other areas in which modal techniques can be applied, some of which are quite well developed and some are still in the initial stages of development. There is also the whole topic of predicate modal logic (of which nothing has appeared in this book). You may be interested in these applications rather than the topics mentioned. Whatever your interests, you will need to know something of the material of this book.

Appendix B

Some solutions to the exercises

B.1 Chapter 1

1.1 (a) Witnessing deductions in a Hilbert systems such as this can be quite long and difficult to find.

(ii) Let

$$
\begin{array}{lll}
\lambda & := & \psi \to \phi \\
\rho & := & \mu \to \nu
\end{array}
\qquad
\begin{array}{lll}
\mu & := & \theta \to \psi \\
\xi & := & \theta \to \lambda
\end{array}
\qquad
\begin{array}{lll}
\nu & := & \theta \to \phi
\end{array}
$$

and then check that

$$
\begin{array}{rcl}
K_1 & : & (\xi \to \rho) \to (\lambda \to (\xi \to \rho)) \\
S_2 & : & \xi \to \rho \\
K_1 S_2 & : & \lambda \to (\xi \to \rho) \\
S_1 & : & (\lambda \to \xi \to \rho) \to (\lambda \to \xi) \to (\lambda \to \rho) \\
S_1(K_1 S_2) & : & (\lambda \to \xi) \to (\lambda \to \rho) \\
K_2 & : & (\lambda \to \xi) \\
(S_1(K_1 S_2))K_2 & : & (\lambda \to \rho)
\end{array}
$$

provides a witnessing deduction. On the left hand side appears a justification for each step, with concatenation indicating a use of MP.

(iii)Let

$$
\begin{array}{lll}
\rho & := & \theta \to \psi \\
\lambda & := & \theta \to (\psi \to \phi)
\end{array}
\qquad
\begin{array}{lll}
\sigma & := & \theta \to \phi \\
\mu & := & \psi \to \rho
\end{array}
\qquad
\begin{array}{lll}
\tau & := & \rho \to \sigma \\
\nu & := & \psi \to \sigma
\end{array}
$$

and

$$
\xi := \mu \to \nu.
$$

By part (ii) we may construct witnessing deduction B_1 and B_2 with

$$
\begin{array}{rcl}
B_1 & : & (\tau \to \xi) \to (\lambda \to \tau) \to (\lambda \to \xi) \\
B_2 & : & \tau \to \xi
\end{array}
$$

and then

$$
\begin{array}{rcl}
B_1 B_2 &:& (\lambda \to \tau) \to (\lambda \to \xi) \\
S_2 &:& \lambda \to \tau \\
B_1 B_2 S_2 &:& \lambda \to \xi \\
S_1 &:& (\lambda \to \xi) \to (\lambda \to \mu) \to (\lambda \to \nu) \\
S_1(B_1 B_2 S_2) &:& (\lambda \to \mu) \to (\lambda \to \nu) \\
K_1 &:& \mu \to (\lambda \to \mu) \\
K_2 &:& \mu \\
K_1 K_2 &:& \lambda \to \mu \\
S_1(B_1 B_2 S_2)(K_1 K_2) &:& \lambda \to \nu
\end{array}
$$

provides the required witnessing deduction.

(iv) An appropriate deduction can be constructed out of the ones given for parts (ii) and (iii).

(b) Using the Deduction Property it suffices to show that

(i) $\phi \vdash \phi$ (ii) $\psi \to \phi, \theta \to \psi, \theta \vdash \phi$

(iii) $\theta \to (\psi \to \phi), \psi, \theta \vdash \phi$ (iv) $\theta \to \psi, \psi \to \phi, \theta \vdash \phi$

(v) $\theta \to (\theta \to \psi), \theta \vdash \phi$

and witnessing deductions for these are easy to find.

1.2 By definition this $\boldsymbol{CON}$ has finite character and, trivially, is basically consistent.

To verify conjunctive preservation consider the case, for instance, where

$$\theta \wedge \psi \in \Phi \in \boldsymbol{CON}.$$

Then each finite subset of

$$\Phi \cup \{\theta, \psi\}$$

is a subset of

$$\Psi \cup \{\theta, \psi\}$$

for some finite $\Psi \subseteq \Phi$. But then $\Psi \cup \{\theta \wedge \psi\}$ is also a finite subset of Φ and hence has a model ν, say. Any model of $\theta \wedge \psi$ is a model of both θ and ψ, so that ν is a model of $\Psi \cup \{\theta, \psi\}$. The other two cases are dealt with in the same way.

To verify disjunctive preservation consider the case, for instance, where

$$\theta \vee \psi \in \Phi \in \boldsymbol{CON}.$$

By way of contradiction suppose that

$$\Phi \cup \{\theta\} \notin \boldsymbol{CON} \quad \text{and} \quad \Phi \cup \{\psi\} \notin \boldsymbol{CON}.$$

This gives us finite subsets Ψ_1, Ψ_2 of Φ such that neither

$$\Psi_1 \cup \{\theta\} \quad \text{nor} \quad \Psi_2 \cup \{\psi\}$$

has a model. Setting $\Psi = \Psi_1 \cup \Psi_2$ we see that neither

$$\Psi \cup \{\theta\} \quad \text{nor} \quad \Psi \cup \{\psi\}$$

has a model. But

$$\Psi \cup \{\theta \vee \psi\}$$

is a finite subset of Φ and hence has a model ν. This valuation models Ψ and at least one of θ or ψ, which leads to the required contradiction.

Negation preservation is proved in a similar fashion.

The compactness result now follows by Theorem 1.5 (for this particular family CON).

1.3 (a) Consider any valuation ν with $[\![\phi]\!]_\nu = 1$. This ν extends some $\pi \in \Pi$, and then

$$[\![\phi^\pi]\!]_\nu \;=\; [\![\phi]\!]_\nu \;=\; 1$$

so that $[\![\lambda]\!] = 1$. Thus $\phi \to \lambda$ is a tautology.

The argument for $\rho \to \psi$ is similar.

Since tautologies are closed under substitution for their variables, we see that

$$\phi^\pi \to \psi^\sigma$$

is a tautology for all $\pi \in \Pi$ and $\sigma \in \Sigma$. Thus $\lambda \to \rho$ follows by several uses of the tautologies

$$(\alpha \to \gamma) \wedge (\beta \to \gamma) \to (\alpha \vee \beta \to \gamma)$$

and

$$(\alpha \to \gamma) \wedge (\alpha \to \delta) \to (\alpha \to \gamma \wedge \delta).$$

(b) Similarly, if

$$\phi \to \theta \quad , \quad \theta \to \psi$$

are tautologies with θ depending only on Q, then so are

$$\phi^\pi \to \theta \quad , \quad \theta \to \psi^\sigma$$

for all π and σ. Hence the required results follow from the above tautologies.

B.2   Chapter 2

2.1 (b)

ϕ	$Pos^+(\phi)$	$Neg^+(\phi)$
$\bot, \top$	Var	Var
P	Var	$\{P\}$
$\neg\theta$	$Neg^+(\theta)$	$Pos^+(\theta)$
$\theta \wedge \psi$	$Pos^+(\theta) \cap Pos^+(\psi)$	$Neg^+(\theta) \cap Neg^+(\psi)$
$\theta \vee \psi$	$Pos^+(\theta) \cap Pos^+(\psi)$	$Neg^+(\theta) \cup Neg^+(\psi)$
$\theta \to \psi$	$Neg^+(\theta) \cap Pos^+(\psi)$	$Pos^+(\theta) \cap Neg^+(\psi)$
$[i]\phi$	$Pos^+(\phi)$	$Neg^+(\theta)$

2.2 Trivially $\phi_1[P := \phi_j]$ is ϕ_j and $\phi_i[P := \phi_1]$ is ϕ_i. Also we have:

$$\phi_2[P := \phi_2] \quad \text{is} \quad \neg\neg P$$
$$\phi_2[P := \phi_3] \quad \text{is} \quad \neg(\neg P \to P)$$
$$\phi_2[P := \phi_4] \quad \text{is} \quad \neg(P \to \neg P)$$
$$\phi_3[P := \phi_2] \quad \text{is} \quad \neg\neg P \to \neg P$$
$$\phi_3[P := \phi_3] \quad \text{is} \quad \neg(\neg P \to P) \to (\neg P \to P)$$
$$\phi_3[P := \phi_4] \quad \text{is} \quad \neg(P \to \neg P) \to (P \to \neg P)$$
$$\phi_4[P := \phi_2] \quad \text{is} \quad \neg P \to \neg\neg P$$
$$\phi_4[P := \phi_3] \quad \text{is} \quad (\neg P \to P) \to \neg(\neg P \to P)$$
$$\phi_4[P := \phi_4] \quad \text{is} \quad (P \to \neg P) \to \neg(P \to \neg P).$$

2.3 (a) We have

$$\xi \ = \ \phi[P := \psi, Q := \theta] \ = \ \psi \to \theta \ = \ Q \to P$$

so that

$$\phi[P := \psi, Q := \theta][Q := \rho, P := \sigma] \ = \ \xi[Q := \rho, P := \sigma] \ = \ \rho \to \sigma.$$

(b) Similarly

$$\lambda \ = \ \psi[Q := \rho, P := \sigma] \ = \ \rho \ , \quad \mu \ = \ \theta[Q := \rho, P := \sigma] \ = \ \sigma$$

so that

$$\phi[P := \lambda, Q := \mu] \ = \ \lambda \to \mu \ = \ \rho \to \sigma$$

as required.

2.4 This is proved by induction on the complexity of ϕ.

2.5 (a) This follows by induction on the complexity of ϕ. The two constant base cases are trivial and the variable case is just the definition of $\tau \bullet \sigma$. The induction steps are equally easy. For instance, when $\phi = \theta * \psi$ for some formulas θ and ψ and binary connective $*$, we have

$$
\begin{aligned}
(\phi^\sigma)^\tau &= ((\theta * \psi)^\sigma)^\tau \\
&= (\theta^\sigma * \psi^\sigma)^\tau \\
&= (\theta^\sigma)^\tau * (\psi^\sigma)^\tau \\
&= \theta^{\tau \bullet \sigma} * \psi^{\tau \bullet \sigma} \quad = \quad (\theta * \psi)^{\tau \bullet \sigma}
\end{aligned}
$$

where the penultimate equality uses the Induction Hypothesis. Similarly for $\phi = [i]\psi$ we have

$$
\begin{aligned}
(\phi^\sigma)^\tau &= (([i]\psi)^\sigma)^\tau \\
&= ([i](\psi^\sigma))^\tau \\
&= [i](\psi^\sigma)^\tau \\
&= [i](\psi^{\tau \bullet \sigma}) \quad = \quad ([i]\psi)^{\tau \bullet \sigma}
\end{aligned}
$$

as required.

(b) For each variable P we have, using (a),

$$
\begin{aligned}
((\tau \bullet \sigma) \bullet \rho)(P) &= (\rho(P))^{\tau \bullet \sigma} \\
&= (\rho(P)^\sigma)^\tau \\
&= ((\sigma \bullet \rho)(P))^\tau \quad = \quad (\tau \bullet (\sigma \bullet \rho))(P)
\end{aligned}
$$

as required.

2.6 (a) The construction tree of $L(P)$ is

$$
\cfrac{\cfrac{\cfrac{\cfrac{P}{\Box P}(\Box) \quad P}{T(P)}(\to)}{\Box T(P)}(\Box) \quad \cfrac{P}{\Box P}(\Box)}{L(P)}(\to)
$$

and then $\Gamma(L(P))$ is the set of formulas occurring at the nodes of this tree. Thus

$$
\Gamma(L(P)) = \{P, \Box P, T(P), \Box T(P), L(P)\}.
$$

(b) This is proved by a simple induction on the complexity of ϕ. For instance, for the case $\phi := \phi_1 * \phi_2$, if

$$
\psi \in \Gamma(\phi) = \Gamma(\phi_1) \cup \Gamma(\phi_2) \cup \{\phi\}
$$

then

$$
\psi \in \Gamma(\phi_1) \quad \text{or} \quad \psi \in \Gamma(\phi_2) \quad \text{or} \quad \psi = \phi
$$

so that the Induction Hypothesis gives

$$
\Gamma(\psi) \subseteq \Gamma(\phi_1) \quad \text{or} \quad \Gamma(\psi) \subseteq \Gamma(\phi_2) \quad \text{or} \quad \Gamma(\psi) \subseteq \Gamma(\phi)
$$

and hence $\Gamma(\psi) \subseteq \Gamma(\phi)$, as required.

B.3   Chapter 3

3.1 To define a transition relation on $A = \{u, v\}$ there are four alternatives to be decided between:

- whether or not $u \longrightarrow u$

- whether or not $u \longrightarrow v$

- whether or not $v \longrightarrow u$

- whether or not $v \longrightarrow v$.

Thus there are 2^4 different such relations as listed.

With $U = \{u\}$, for each $a \in A$ we have $a \in \Box U$ precisely when

$$(\forall x \in A)[a \longrightarrow x \;\Rightarrow\; x = u] \quad \text{i.e.} \quad (\forall x \in A)[x \neq u \;\Rightarrow\; \neg(a \longrightarrow x)]$$

so that

$$a \in \Box U \;\Leftrightarrow\; \neg[a \longrightarrow v].$$

This enables us to compute $\Box U$, and similar arguments enable us to compute $\Box V$ and $\Box \emptyset$.

3.2 With $A = \{u, v, w\}$ there are 9 ordered pairs of elements taken from A. Each of these pairs may or may not be part of a transition relation so there are 2^9 different transition relations on A.

3.3 These follow from the corresponding results for $\Box$. Thus

$$
\begin{aligned}
X \subseteq Y \;\Rightarrow\;& \neg Y \subseteq \neg X \\
\Rightarrow\;& \Box\neg Y \subseteq \Box\neg X \\
\Rightarrow\;& \neg\Box\neg Y \subseteq \neg\Box\neg X \;\Rightarrow\; \Diamond X \subseteq \Diamond Y.
\end{aligned}
$$

Similarly

$$\Diamond\emptyset = \neg\Box\neg\emptyset = \neg\Box A = \neg A = \emptyset$$

and

$$
\begin{aligned}
\Diamond(X \cup Y) &= \neg\Box\neg(X \cup Y) \\
&= \neg\Box(\neg X \cap \neg Y) \\
&= \neg(\Box\neg X \cap \Box\neg Y) \\
&= \neg\Box\neg X \cup \neg\Box\neg Y \;=\; \Diamond X \cup \Diamond Y
\end{aligned}
$$

as required.

3.4 (a)(i) Since the relation $\longrightarrow$ is reflexive we have $\Box X \subseteq X$. In particular $\Box\neg X \subseteq \neg X$ so that $X \subseteq \neg\Box\neg X = \Diamond X$. This gives

$$\Box X \subseteq X \subseteq \Diamond X$$

and particular cases of these give

$$\Box X \subseteq \Diamond \Box X \quad , \quad \Box \Diamond X \subseteq \Diamond X.$$

Now consider any $a \in \Diamond \Box X$. Then there is some $b \in A$ with $a \longrightarrow b$ and $b \in \Box X$, i.e.

$$b \longrightarrow x \Rightarrow x \in X$$

for all $x \in A$. Since $\longrightarrow$ is symmetric we may set $x = a$ to get $a \in X$. Thus we have $\Diamond \Box X \subseteq X$ and a dual argument gives $X \subseteq \Box \Diamond X$.

(ii) We now have

$$\Diamond \Box X \subseteq X \subseteq \Box \Diamond X.$$

The monotonicity of $\Box$ on the first inclusion gives $\Box \Diamond \Box X \subseteq \Box X$ and a particular case of the second inclusion gives $\Box X \subseteq \Box \Diamond \Box X$, hence the required equality.

(b) The operation $\Box$ is idempotent precisely when it is nearly inflationary, i.e. when $\longrightarrow$ is transitive. There are reflexive, symmetric relations which are not transitive.

3.5 (i) Suppose $a \in [\bullet]X$. Then

$$a \longrightarrow x \Rightarrow x \in X$$

holds for all $x \in A$, so that $a \in \Box X$.

(ii) An easy computation shows that

$$\Box \bigcap \mathcal{X} = \bigcap \Box \mathcal{X}$$

for all $\mathcal{X} \subseteq \mathcal{P}A$. Hence if $\Box = [\bullet]$ then $[\bullet]$ also has this intersection preserving property.

Conversely, suppose that $[\bullet]$ has this property. For each $X \in \mathcal{P}A$ we have

$$X = \bigcap \{ \neg\{y\} \mid y \in \neg X \}$$

so that

$$[\bullet]X = \bigcap \{ [\bullet]\neg\{y\} \mid y \in \neg X \}.$$

Now suppose that $a \in \Box X$. Then for each $y \in \neg X$ we have $\neg[a \longrightarrow y]$ so there is some $Z \in \mathcal{P}A$ with

$$a \in [\bullet]Z \quad , \quad y \in \neg Z.$$

Using the previous observation (with Z for X) this gives some $a \in [\bullet]\neg\{y\}$ and hence

$$y \in \neg X \Rightarrow a \in [\bullet]\neg\{y\}.$$

Thus

$$a \in \bigcap \{ [\bullet]\neg\{y\} \mid y \in \neg X \} = [\bullet]X$$

which is enough to show that $\square = [\bullet]$.

(iii) Let R be the set of real numbers and consider the operation $[\bullet]$ on $\mathcal{P}$R given by

$$a \in [\bullet]X \Leftrightarrow (\exists l, r \in \mathsf{R})[a \in (l, r) \subseteq X].$$

(This is the metric interior operation.) In particular

$$[\bullet][0, 1] = (0, 1).$$

We easily check that the induced relation $\longrightarrow$ is just equality, and hence $\square X = X$ for all $X \in \mathsf{R}$.

B.4   Chapter 4

4.1 The completed table is as follows.

	D	T	B	4	5	P	Q	R	G	L	M
(1)	×	×	×	✓	✓	×	✓	✓	✓	×	×
(2)	✓	✓	✓	✓	×	✓	×	✓	✓	×	×
(4)	×	✓	✓	×	✓	✓	✓	×	×	×	✓
(5)	×	×	✓	✓	×	×	×	✓	×	×	×
(7)	×	✓	×	✓	×	✓	×	✓	×	×	×
(8)	×	×	×	×	✓	×	×	✓	×	×	✓
(11)	✓	✓	✓	✓	✓	×	×	✓	×	×	✓
(13)	✓	✓	✓	✓	✓	×	✓	✓	×	✓	✓
(14)	×	×	✓	×	×	✓	×	✓	×	×	×
(16)	✓	✓	✓	✓	×	×	×	×	×	×	×

These entries can be verified by rather tedious arguments. However, all the columns except M can be verified much more easily using the results of correspondence theory which is the subject of Chapter 5. I suggest you wait until you have learned this technique before you do too many of the computations.

To deal with the shape M, note that a structure $\mathcal{A}$ models M(ϕ) if and only if

$$a \Vdash \Diamond \square \phi \vee \Diamond \square \neg \phi \qquad (\text{B.1})$$

for each $a \in A$ and valuation on $\mathcal{A}$. In particular, if $\mathcal{A}$ does model M then no element is blind. (See Exercise 4.6 for this notion.) Thus none of (1),(2),(5), and (7) models M.

To show that the structures (4),(8),(11), and (13) do model M show (by considering each case separately) that for each element a there is some element b such that

$$a \Vdash \Diamond \square \phi \Leftrightarrow b \Vdash \phi$$

(for all formulas ϕ and valuations on the structure). The table

	$a = u$	$a = v$
4	u	v
8	v	v
11	u	v
13	u	u

gives the appropriate elements b for a.

Finally, for (14) and (16), check that

$$u \Vdash \Diamond \Box \phi \iff u \Vdash \phi \text{ and } v \Vdash \phi$$

which (by making ϕ true at u and false at v) is enough to show that M is not modelled by these structures.

4.2 Consider the two structures

$$u \qquad\qquad v \qquad\qquad w$$

$$\odot \longleftrightarrow \circ \longrightarrow \circ$$

where u may or may not be reflexive, but v and w are definitely not reflexive.

4.3

	D	T	B	4	5	P	Q	R	G	L	M
(a)	✓	×	×	✓	×	×	×	×	✓	×	×
(b)	✓	✓	×	✓	×	×	×	✓	✓	×	×
(c)	×	×	×	✓	×	×	×	×	×	✓	✓
(d)	✓	✓	×	✓	×	×	×	✓	✓	×	✓

4.4 (i - iv) For $\mathcal{A} = \mathcal{N}, \mathcal{Z}$, any valuation on $\mathcal{A}$, any $a \in A$, and any formula ϕ, suppose that

$$a \Vdash \Diamond \Box \phi \quad , \quad a \Vdash \Box X(\phi)$$

where $X(\phi)$ is

$$\Box \phi \to \phi \quad \text{or} \quad \Box(\phi \to \Box \phi) \to \phi$$

as appropriate. We require that $a \Vdash \Box \phi$.

There is some $b > a$ with $b \Vdash \Box \phi$. We take the least such b. Note that $b \Vdash X(\phi)$, and a simple argument shows that

$$b \Vdash \Box \phi \quad \text{and} \quad b \Vdash \Box(\phi \to \Box \phi)$$

so that $b \Vdash \phi$. This gives $b - 1 \Vdash \Box \phi$ so that (by the minimality of b) we have $a = b - 1$, and hence $a \Vdash \Box \phi$, as required.

(v - viii) For $\mathcal{A} = \mathcal{Q}, \mathcal{R}$ consider any subset $D \subseteq A$ such that for each $a, b \in A$ with $a < b$, there are $d \in D$ and $e \in A - D$ with

$$a < d < b \quad , \quad a < e < b.$$

(For instance, D may be the dyadic rationals.) Consider any valuation on $\mathcal{A}$ where

$$x \Vdash P \Leftrightarrow x \in D \text{ or } 1 \le x$$

(for $x \in A$). Clearly

$$0 \Vdash \Diamond \Box P \quad , \quad 0 \Vdash \neg \Box P$$

so it suffices to show that

$$0 \Vdash \Box T(P) \quad , \quad 0 \Vdash \Box U(P).$$

To this end we show that

$$a \Vdash T(P) \quad , \quad a \Vdash U(P)$$

for all $a \in A$.

Consider any $a \in A$ with $a \Vdash \neg P$. Then $a \in A - D$ and $a < 1$. Take any $e \in A - D$ with $a < e < 1$. Then $e \Vdash \neg P$, so that $a \Vdash \neg \Box P$, and hence $a \Vdash T(P)$. Similarly, take any $d \in D$ with $a < d < e$. Then $d \Vdash P \wedge \neg \Box P$ so that $a \Vdash \neg \Box (P \to \Box P)$, and hence $a \Vdash U(P)$.

4.5 The implication

$$\Diamond^2 \phi \to \Diamond \phi$$

is universally valid in all transitive structures, hence so is

$$\Box \Diamond^2 \phi \to \Box \Diamond \phi.$$

Conversely suppose that

$$a \Vdash \Box \Diamond \phi$$

(for some $a \in A$, valuation on $\mathcal{A}$, and formula ϕ) and consider any $a \longrightarrow b$. We require that $b \Vdash \Diamond^2 \phi$. But the position of a gives $b \Vdash \Diamond \phi$, which provides some c with

$$a \longrightarrow b \longrightarrow c.$$

Transitivity gives $a \longrightarrow c$ so that $c \Vdash \Diamond \phi$ and hence there is some d with

$$a \longrightarrow b \longrightarrow c \longrightarrow d \quad , \quad d \Vdash \phi.$$

Transitivity now gives $b \longrightarrow d$ so that $b \Vdash \Diamond \phi$, as required.

The second required equivalence follows by a similar, but more involved, argument.

4.6 (c)

(i) $\Box \Box \bot$ (ii) $\Diamond \Diamond \top$

(iii) $\Box (\Box \bot \vee \Diamond \Diamond \top)$ (iv) $\Diamond (\Diamond \top \wedge \Box \Box \bot)$

4.7 (b) If $a \xrightarrow{l+1} x$ for some $x \in A$, then either

$$x \xrightarrow{k} b \ \text{ or } \ x \xrightarrow{k} c$$

for some $k \neq 0$, hence $x \Vdash \Diamond \top$. There is no $x \in A$ with $a \xrightarrow{n+2} x$.

(c) The element c witnesses that

$$a \Vdash \Diamond^{m+2} \top$$

whereas b witnesses that

$$a \Vdash \neg \square^{m+1} \Diamond \top$$

and hence $a \Vdash \neg W(m)$. By (b), for each $k \neq m$, either

$$a \Vdash \square^{k+2} \bot \ \text{ or } \ a \Vdash \square^{k+1} \Diamond \top$$

hence $a \Vdash W(k)$.

4.8 (b) Only the shapes D,T,R, and G are modelled by $\mathcal{N}$.

(c) (i) Modelled by $\mathcal{N}$.

(ii) Modelled by $\mathcal{N}$. For instance, consider $X \subseteq \mathsf{N}$ of type (2) with the given a. Then we may check that

$$
\begin{aligned}
\square X &= [a+2, \infty] \\
\square^2 X &= [a+3, \infty] \\
\square^3 X &= [a+4, \infty] \\
\square X \to \square^2 X &= \{a+2\}' \\
\square^2 X \to \square^3 X &= \{a+3\}' \\
\square(\square^2 X \to \square^3 X) &= [a+3, \infty] \subseteq \{a+2\}'
\end{aligned}
$$

hence the required result.

(iii) If X is neither finite nor cofinite then

$$\square X = \square^2 X = \square^3 X = \square^4 X = \emptyset$$

and

$$\square(\square(X \to \square X) \to \square^3) = \square \Diamond X = \square \mathsf{N} = \mathsf{N}$$

which shows that the formula is not modelled by $\mathcal{N}$.

4.9 (a)(ii) For a given valued structure $(\mathcal{A}, \alpha)$ and element $a \in A$, let

$$B = A(a)$$

be the set of all elements $b \in A$ for which $a \xrightarrow{i} b$ for some compound label i, i.e. for which there is a chain

$$a = a_0 \xrightarrow{i(1)} a_1 \xrightarrow{i(2)} \cdots \xrightarrow{i(n)} a_n = b$$

for some labels $i(1), i(2), \ldots, i(n)$. Let $\mathcal{B} = \mathcal{A}(a)$ be the substructure of $\mathcal{A}$ based on $A(a)$. Let β be the restriction of α to $\mathcal{B}$.

(b) This is proved by induction on ϕ. Only the induction step across modal operators is non-trivial. The equivalence

$$(\mathcal{B}, \beta, b) \Vdash [i]\phi \Leftrightarrow (\mathcal{A}, \alpha, b) \Vdash \phi$$

is proved as follows. Suppose the LHS holds and consider any $a \in A$ with $b \xrightarrow{i} a$. Then, by genericity, $a \in B$ so that $(\mathcal{B}, \beta, b) \Vdash \phi$ and we may use the Induction Hypothesis. This gives the implication $\Rightarrow$ and the converse implication is trivial.

(c) Suppose first that

$$(\mathcal{A}, \alpha, a) \Vdash^p \{\phi\}^* \quad \text{i.e. that} \quad (\mathcal{A}, \alpha, a) \Vdash^p [i]\phi$$

for all compound labels i. For each element b of $\mathcal{B}$ we have $a \xrightarrow{i} b$ for some compound label i so that

$$(\mathcal{A}, \alpha, b) \Vdash^p \phi$$

and hence part (b) gives $(\mathcal{B}, \beta, b) \Vdash^p \phi$. Since this holds for arbitrary elements b of $\mathcal{B}$, this gives $(\mathcal{B}, \beta) \Vdash^v \phi$, as required.

The required converse follows by a similar, but dual, argument.

4.10 (a) Unravelling the definition we see that

$$a^\vee \xrightarrow{i} b^\vee \tag{B.2}$$

holds precisely when

$$(\forall X \in \mathcal{P}A)[a \in [i]X \Rightarrow b \in X]. \tag{B.3}$$

Recall also that

$$a \in [i]X \Leftrightarrow (\forall x \in A)[a \xrightarrow{i} x \Rightarrow x \in X]. \tag{B.4}$$

Suppose that (B.2) holds and consider the set

$$X = \{x \in A \mid a \xrightarrow{i} x\}.$$

Then $a \in [i]x$, so that $b \in X$, and hence $a \xrightarrow{i} b$. Conversely, if $a \xrightarrow{i} b$, then (B.3) follows by taking $x = b$ in (B.4).

(b) This is proved by induction on the complexity of ϕ. Only the induction step

$$p \Vdash [i]\phi \Leftrightarrow [\![\,[i]\phi\,]\!] \in p$$

is non-trivial, and even here the implication $\Leftarrow$ follows immediately from the definition of $\xrightarrow{i}$ in $\mathcal{A}^\vee$. For the converse, consider the family

$$\{X \in \mathcal{P}A \mid [i]X \in p\} \cup \{[\![\phi]\!]'\}.$$

If this has the f.i.p. then it extends to some $q \in A^\vee$ which then witnesses that $p \Vdash \langle i \rangle \neg \phi$. If it does not have the f.i.p. then there is some $X \in \mathcal{P}A$ with

$$[i]X \in p \quad , \quad X \cap [\![\phi]\!]' = \emptyset.$$

The second of these gives $X \subseteq [\![\phi]\!]$ so that

$$[i]X \subseteq [i][\![\phi]\!] = [\![[i]\phi]\!]$$

and hence $[\![[i]\phi]\!] \in p$.

(c) (i) Consider the case $p = a^\vee$ in (b).

(ii) The implication $\Rightarrow$ holds by (i). Conversely, if $\neg[(\mathcal{A}, \alpha)^\vee \Vdash^v \phi]$ then, by (b), there is some $p \in A^\vee$ with $[\![\phi]\!] \notin p$. This gives $[\![\phi]\!] \ne A$.

(iii) By (i).

4.11 (a) Observe that

$$a \in \square X \iff [a+1, \infty] \subseteq X$$

and recall that no member of $A^\vee$ contains $\emptyset$.

(b) Each non-principal ultrafilter contains all cofinite sets.

(c) Each non-principal ultrafilter is a reflexive point.

(d) There are many formulas ϕ such that for all transitive structures $\mathcal{A}$ we have

$$\mathcal{A} \Vdash^u \phi \iff \mathcal{A} \text{ has no reflexive points}$$

and many of these are modelled by $\mathcal{N}$. One such formula is $L(P)$ (for any variable P).

4.12 (a) Trivially, $\mathcal{N}$ is transitive and serial but has no reflexive elements.

An easy argument shows that $\mathcal{N}^\vee$ is transitive. Also, by Exercise 4.11, for each $p, q \in N^\vee$ with q non-principal, we have

$$p \longrightarrow q \longrightarrow q$$

which shows that $\mathcal{N}^\vee$ is good.

(b) If Γ captures any class of good structures which contains $\mathcal{N}^\vee$, then $\mathcal{N}^\vee \Vdash^u \Gamma$ so that, by Exercise 4.10(c)(iii), we have $\mathcal{N} \Vdash^u \Gamma$, which would mean that $\mathcal{N}$ is good.

B.5   Chapter 5

5.1 For all cases the implication

$$\mathcal{A} \text{ has (p)} \ \Rightarrow\ \mathcal{A} \text{ models (s)}$$

is straight forward. For the converse, replace ϕ by a variable P, fix the elements a and b with $a \xrightarrow{i} b$, and consider a valuation β with $\beta(P) = \{b\}$. For instance case (g) is proved as follows.

Suppose first that $\mathcal{A}$ has (p) and, for an arbitrary valuation α and element a, that $a \Vdash \langle i \rangle \phi$. This gives some b with $a \xrightarrow{i} b$ and $b \Vdash \phi$. We require that

$$a \Vdash [j]\,\langle k \rangle\,[l]\phi.$$

To this end consider any c with $a \xrightarrow{j} c$. This wedge with (p) gives a particular d with

$$d \xrightarrow{l} x \ \Rightarrow\ x = b \ \Rightarrow\ x \Vdash \phi$$

so that $d \Vdash [l]\phi$. Thus $c \Vdash \langle k \rangle\,[l]\phi$ which gives the required result.

Conversely, suppose that $\mathcal{A}$ models

$$\langle i \rangle P \to [j]\,\langle k \rangle\,[l]p.$$

For a given wedge

$$
\begin{array}{ccc}
a & \xrightarrow{\ \ i\ \ } & b \\[2pt]
{\scriptstyle j}\big\downarrow & & \\[2pt]
c & &
\end{array}
$$

consider the valuation β as above. Then $a \Vdash \langle i \rangle P$ so that $a \Vdash [j]\,\langle k \rangle\,[l]P$ and hence $c \Vdash \langle k \rangle\,[l]P$. This provides the required element d.

5.2 For all cases the implication

$$\mathcal{A} \text{ has (p)} \ \Rightarrow\ \mathcal{A} \text{ models (s)}$$

is straight forward. For the converse suppose that $\mathcal{A}$ models

$$\langle i \rangle\,[j]P \to \mathrm{RHS}(P)$$

for some variable P, and for a given transition

$$a \xrightarrow{i} b$$

consider any valuation on $\mathcal{A}$ such that

$$x \Vdash P \ \Leftrightarrow\ b \xrightarrow{j} x$$

(for all $x \in A$). Then $a \Vdash \langle i \rangle [j]P$ so that $a \Vdash \mathrm{RHS}(P)$ which, by a simple argument, gives the required result.

5.3 Assume the structural property and suppose first that for some valuation and elements a, b, c we have

$$a \xrightarrow{l} b \xrightarrow{n} c$$

and

$$a \Vdash [i]([j]\phi \to [k]\psi) \quad , \quad b \Vdash [m]\phi.$$

The structural property supplies an element d for which we may check that

$$d \Vdash [j]\phi \to [k]\psi \quad , \quad d \Vdash [j]\phi.$$

and hence get $c \Vdash \psi$.

Secondly, replace ϕ and ψ by distinct variables P and Q and suppose $\mathcal{A}$ models this formula. For given elements a, b, c consider any valuation such that

$$x \Vdash P \Leftrightarrow b \xrightarrow{m} x \quad , \quad x \Vdash Q \Leftrightarrow x \neq c.$$

Observe that

$$b \Vdash \neg([m]P \to [n]Q)$$

so that

$$a \Vdash \neg[i]([m]P \to [n]Q)$$

which produces the required element d.

5.4 Property (ii) follows from (i) by setting $\phi := \top$ and remembering that $\Box \top$ is universally valid. Property (iii) is just a reformulation of (ii).

To show that (iii) $\Rightarrow$ (i) consider any $a \in A$ and formula ϕ with

$$a \Vdash \Box^k \Diamond \Box \phi$$

(for some valuation). Then, for each b with $a \xrightarrow{k} b$ we have $b \Vdash \Diamond \Box \phi$, in particular no such b is blind. Thus, using (iii), we see there is also no blind b with $a \xrightarrow{l} b$. Consider any such b. We must show that $b \Vdash \Diamond \Box \phi$. Since b is not blind, there is at least one x with $b \longrightarrow x$, furthermore (using transitivity)

$$b \longrightarrow x \Rightarrow a \xrightarrow{l} x \Rightarrow x \text{ is not blind.}$$

By iteration this allows us to obtain some c with

$$a \xrightarrow{l} b \longrightarrow c \quad , \quad a \xrightarrow{k} c$$

and hence $c \Vdash \Diamond \Box \phi$. This gives some d with

$$b \longrightarrow c \longrightarrow d \quad , \quad d \Vdash \Box \phi$$

which, again using transitivity, gives $b \Vdash \Diamond \Box \phi$, as required.

5.5 The implication (i) $\Rightarrow$ (ii) holds since $\Diamond \Box \phi \to \Diamond \phi$ is universally valid in all transitive structures. Condition (iii) follows from (ii) by setting $\phi := \mathsf{T}$, and (iv) is a simple reformulation of (iii).

To show (iv) $\Rightarrow$ (i) consider any $a \in A$ and formula ϕ with $a \Vdash \Box^k \Diamond \phi$ (for some valuation). In particular we have

$$a \xrightarrow{k} x \Rightarrow x \text{ is not blind}$$

for each $x \in A$. Thus (iv) gives some $b \in A$ with $a \xrightarrow{l} b$ and

$$b \xrightarrow{k} x \Rightarrow x \text{ is not blind}.$$

With this b we show that $b \Vdash \Box \Diamond \phi$.

To this end consider any c with $b \longrightarrow c$, This c is not blind and, by transitivity,

$$c \longrightarrow x \Rightarrow b \longrightarrow x \Rightarrow x \text{ is not blind}.$$

Thus, by iteration, we obtain some d with

$$a \xrightarrow{l} b \longrightarrow c \longrightarrow d \quad , \quad a \xrightarrow{k} d.$$

But then $d \Vdash \Diamond \phi$ hence, by transitivity, $c \Vdash \phi$, which gives the required result.

5.6 (b) This follows easily by induction on M. For instance

$$a -\{ [i]\mathrm{M}\}\!\!\rightarrow b$$

holds precisely when, for each x,

$$a \xrightarrow{i} x \Rightarrow x -\{\mathrm{M}\}\!\!\rightarrow b$$

which, by the Induction Hypothesis, is equivalent to

$$a \xrightarrow{i} x \Rightarrow x \Vdash \mathrm{M}P$$

i.e. when $a \Vdash [i]\mathrm{M}P$.

(c) Fix a valuation and element b and for each modal operator M consider the property

((M))   For each element a with $a -\{\mathrm{M}\}\!\!\rightarrow b$,

$$b \Vdash \phi \Rightarrow a \Vdash \mathrm{M}\phi$$

holds.

We verify ((M)) by induction on M. For instance, to prove the Induction Step

$$((M)) \Rightarrow ((\,[i]M))$$

consider an element a with $a -\{\,[i]M\}\!\!\rightarrowtail b$ and suppose that $b \Vdash \phi$ for some formula ϕ. We require that $a \Vdash [i]M\phi$. To this end consider any element x with $a \overset{i}{\longrightarrow} x$. Then $x -\{M\}\!\!\rightarrow b$ and so the Induction Hypothesis gives $x \Vdash M\phi$, which is the desired result.

(d) Suppose first that

$$\overset{i}{\longrightarrow} \;\subseteq\; -\{M\}\!\!\rightarrow$$

and that $a \Vdash \langle i\rangle\phi$ (for some formula ϕ, valuation on $\mathcal{A}$ and element a). This gives us some element b with $a \overset{i}{\longrightarrow} b$ and $b \Vdash \phi$. The structural property now gives $a -\{M\}\!\!\rightarrow b$ so that, by (c), we have $a \Vdash M\phi$, as required.

Conversely, suppose that $\mathcal{A}$ models

$$\langle i\rangle P \to MP$$

(for some variable P), and consider any pair $a \overset{i}{\longrightarrow} b$. Using a valuation β with $\beta(P) = \{b\}$, we have $a \Vdash \langle i\rangle P$, so that $a \Vdash MP$, and hence (b) gives $a -\{M\}\!\!\rightarrow b$, as required.

(e) For instance

$$\overset{i}{\longrightarrow} \;\subseteq\; -\{\,[j]\,\langle k\rangle\,[l]\,\}\!\!\rightarrow$$

holds precisely when for each wedge

we have $c -\{\,\langle k\rangle\,[l]\,\}\!\!\rightarrowtail b$, i.e. when there is some d with

$$c \overset{k}{\longrightarrow} d -\{\,[l]\,\}\!\!\rightarrowtail b$$

as required.

5.7 (b) (i, ii) Since

$$(x,i) \longrightarrow u \Leftrightarrow u = (x,i)$$

holds for all $x \in S, i \in 2$, and $u \in A(F)$.

(iii) If $x \Vdash \Diamond\,\Box\phi$ then $(x,i) \Vdash \Diamond\phi$ for both $i = 0,1$, so that $(x,i) \Vdash \phi$ and hence $(x,i) \Vdash \Box\phi$.

(iv) By tightening the argument of (iii).

(v) If $f \Vdash \Box\,\Diamond\phi$ then $(x, f(x)) \Vdash \Diamond\phi$ for all $x \in S$, so that $(x, f(x)) \Vdash \Box\phi$ for each such x, and hence $f \Vdash \Diamond\,\Box\phi$.

(vi) By tightening the argument of (v).

(c) (i) For each $x \in S$ we have $(x, g(x)) \Vdash P$ so that $x \Vdash \Diamond P$. For each $f \in F$ we have $f \neq \neg g$ which gives some $x \in S$ with

$$f(x) = g(x)$$

and hence $f \Vdash \Diamond P$. Thus $a \Vdash \Box \Diamond P$.

(ii) If $\mathcal{A}$ models $\mathrm{M}(P)$ then $a \Vdash \Diamond \Box P$ so that either

$$x \Vdash \Box P \quad \text{or} \quad f \Vdash \Box P$$

for some $x \in S$ or $f \in F$. The first of these gives $g(x) = i$ for both $i = 0, 1$ which is impossible. The second gives $f(x) = g(x)$ for each $x \in S$, i.e. $g = f \in F$ as required.

(d) Suppose first that $F = [S \longrightarrow 2]$. It suffices to show that $a \Vdash \mathrm{M}(\phi)$. Thus suppose that $a \Vdash \Box \Diamond \phi$. This gives

$$x \Vdash \Diamond \phi \quad , \quad F \Vdash \Diamond \phi$$

for all $x \in S$ and $f \in F$. The first of these gives some $f \in F$ with

$$(x, f(x)) \Vdash \phi$$

for all $x \in S$, and hence $f \Vdash \Box \phi$. Thus $a \Vdash \Diamond \Box \phi$, as required.

For the converse, if $\mathcal{A}(F)$ models $\mathrm{M}(P)$ then the $\neg(\cdot)$-closure with part (c) gives, for each $g \in [S \longrightarrow 2]$

$$g \notin F \Rightarrow \neg g \notin F \quad \text{and} \quad \neg g \notin F \Rightarrow g \in F$$

i.e. $g \in F$.

5.8 (a) Suppose first that $\mathcal{A}$ has the structural property and that $a \Vdash \Box \Diamond \phi$ (for some element a, valuation, and formula ϕ). The property gives us some b with $a \longrightarrow b$ and

$$b \Vdash \Diamond \phi \quad \text{and} \quad b \longrightarrow x \Rightarrow x = b$$

(for all $x \in A$). These give

$$b \Vdash \phi \quad \text{and} \quad b \longrightarrow x \Rightarrow x \Vdash \phi$$

so that $b \Vdash \Box \phi$, and hence $a \Vdash \Diamond \Box \phi$. (Note that this does not use the transitivity of the structure $\mathcal{A}$.)

Conversely, suppose that $\mathcal{A}$ models

$$\Box \Diamond P \rightarrow \Diamond \Box P$$

for some variable P. Given any $a \in A$ let

$$X = \{x \in A \mid a \longrightarrow x\}.$$

We wish to show that

$$(\exists x \in X)(\forall y \in X)[x \longrightarrow y \;\Rightarrow\; x = y].$$

If this doesn't hold then, using the given choice principle $(*)$, we obtain a certain partition Y, Z of X. Consider any valuation on $\mathcal{A}$ such that for each $x \in X$

$$x \Vdash P \;\Leftrightarrow\; x \in Y \quad , \quad x \Vdash \neg P \;\Leftrightarrow\; x \in Z.$$

Then

$$x \Vdash \Diamond P \wedge \Diamond \neg P$$

for each $x \in X$, so that

$$a \Vdash \Box \Diamond P \wedge \Box \Diamond \neg P$$

which contradicts the given McKinsey axiom.

(b) Consider the set $\mathcal{P}$ of all disjoint pairs $Y, Z \subseteq X$ satisfying

$$(\forall y \in Y)(\exists z \in Z)[y \longrightarrow z] \quad , \quad (\forall z \in Z)(\exists y \in Y)[z \longrightarrow y].$$

$\mathcal{P}$ is non-empty since the pair $(\emptyset, \emptyset) \in \mathcal{P}$. Consider any pair $(Y, Z) \in \mathcal{P}$ and suppose there is some $a \in X - (Y \cup Z)$. The given property of the relation $\longrightarrow$ on X ensures the existence of a sequence

$$a = a_0 \longrightarrow a_1 \longrightarrow a_2 \longrightarrow \cdots \longrightarrow a_r \longrightarrow \cdots$$

where $a_r \neq a_{r+1}$ for all appropriate r. Note also that $a_r \longrightarrow a_s$ whenever $r < s$. We continue this sequence indefinitely unless either

repeat there is some n such that a_{n+1} is an earlier term,

or

capture there is some n such that $a_{n+1} \in Y \cup Z$.

If either of these occur, we take the first such n and consider only

$$a_0, a_1, \ldots, a_n.$$

Now split this sequence into two parts

$$U \; :- \; a_0, a_2, a_4, \ldots \quad , \quad V \; :- \; a_1, a_3, a_5, \ldots$$

and set either

$$Y^+ = Y \cup U \quad , \quad Z^+ = Z \cup V$$

or

$$Y^+ = Y \cup V \quad , \quad Z^+ = Z \cup U.$$

One of these pairs is in $\mathcal{P}$. Only in the 'capture' case do we have to be careful about which pair we take. In this case we note the parity of n and whether $a_{n+1} \in Y$ or $a_{n+1} \in Z$.

The set $\mathcal{P}$ is partially ordered by pairwise inclusion. By the Axiom of Choice there is a maximal pair in $\mathcal{P}$. The above construction shows that such a maximal pair covers X. is good.

B.6 Chapter 6

6.1 In each case there are two possible formulas where one is the contrapositive of the other.

(a) $\langle i \rangle [j] \phi \to \langle i \rangle \langle j \rangle \phi$ $[i][j]\phi \to [i]\langle j \rangle \phi$

(b) $\langle i \rangle [j] \phi \to \langle k \rangle \phi$ $[k]\phi \to [i]\langle j \rangle \phi$

(c) $[i][j]\phi \to \phi$ $\phi \to \langle i \rangle \langle j \rangle \phi$

(d) $\langle i \rangle \langle j \rangle [k]\phi \to \phi$ $\phi \to [i][j]\langle k \rangle \phi$

(e) $\langle i \rangle \langle j \rangle \phi \to \langle k \rangle \phi$ $[k]\phi \to [i][j]\phi$

(f) $\langle i \rangle \langle j \rangle \langle k \rangle \phi \to \langle l \rangle \langle m \rangle \phi$ $[l][m]\phi \to [i][j][k]\phi$

6.2 (a) $[i](\langle j \rangle [l]\phi \to [k]\langle m \rangle \phi)$

(b) $\langle i \rangle [k]\phi \to [j]\langle l \rangle(\phi \wedge \langle m \rangle \top)$

(c) $(\langle i \rangle [l]\phi \wedge \langle j \rangle [m]\psi) \to \langle k \rangle(\phi \wedge \psi)$

(d) $(\langle i \rangle [l]\phi \wedge \langle k \rangle [n]\psi) \to [j]\langle m \rangle(\phi \wedge \psi)$

6.3 The appropriate structural property is as follows.

For all elements $a, b, c(1), c(2), \ldots, d$ with

$$a \xrightarrow{i} b \xrightarrow{l} d \quad \text{and} \quad b \xrightarrow{j(p)} c(p) \tag{B.5}$$

for $p = 1, 2, \ldots$, there are elements e and f with

$$d \xrightarrow{m} e \xrightarrow{n} f \quad \text{and} \quad c(p) \xrightarrow{k(p)} e \xrightarrow{n} f$$

for $p = 1, 2, \ldots$.

Showing that this property ensures the modelling of the axiom is straight forward. Conversely, suppose that $P_1, P_2, \ldots$ are distinct variables and suppose that the structure $\mathcal{A}$ models

$$[i](\bigwedge\{\langle p \rangle [p]P_p \mid p = 1, 2, \ldots\} \to [l]\langle m \rangle(\langle n \rangle \top \wedge \bigwedge\{P_p \mid p = 1, 2, \ldots\}))$$

(which, of course, is an instance of the axiom). Suppose also that we are given elements $a, b, c(1), c(2), \ldots, d$ of $\mathcal{A}$ satisfying (B.5). We may define a valuation on $\mathcal{A}$ such that

$$x \Vdash P_p \Leftrightarrow c(p) \xrightarrow{k(p)} x$$

for all $x \in A$ and $p = 1, 2, \ldots$. In particular

$$c(p) \Vdash [k]P_p \quad \text{where } k = k(p)$$

and hence

$$b \Vdash \bigwedge \{ \langle p \rangle [p]P_p \mid p = 1, 2, \ldots \}.$$

The instance of the axiom now gives

$$b \Vdash [l] \langle m \rangle (\langle n \rangle \top \wedge \bigwedge \{ P_p \mid p = 1, 2, \ldots \})$$

and hence

$$d \Vdash \langle m \rangle (\langle n \rangle \top \wedge \bigwedge \{ P_p \mid p = 1, 2, \ldots \}).$$

This produces an element e with

$$d \xrightarrow{m} e \quad , \quad e \Vdash \langle n \rangle \top \quad , \quad e \Vdash P_p$$

which leads to the required result.

B.7   Chapter 7

7.2 (iii)$\Rightarrow$(i) Suppose that (iii) holds and consider any valued pointed model $(\mathcal{A}, \alpha, a)$ of Φ^*. Let $\mathcal{B} = \mathcal{A}(a)$ be the substructure generated by a and let β be the restriction of α to $\mathcal{B}$. By Exercise (4.9(c)) we have $(\mathcal{B}, \beta) \Vdash^v \Phi$ and hence, by (iii), we have $(\mathcal{B}, \beta) \Vdash^v \phi$, which gives $(\mathcal{B}, \beta, b) \Vdash^p \phi$ and hence $(\mathcal{A}, \alpha, a) \Vdash^p \phi$, as required.

B.8   Chapter 8

8.1 (i) Since

$$\theta \to (\psi \to \theta \wedge \psi)$$

is an instance of a tautology, use of (EN) and (K) give

$$\vdash \Box \theta \to (\Box \psi \to \Box(\theta \wedge \psi))$$

and hence

$$\vdash (\Box \theta \wedge \Box \psi) \to \Box(\theta \wedge \psi).$$

Also $\theta \wedge \psi \to \theta$ is an instance of a tautology so that (EN) gives

$$\vdash \ \Box(\theta \wedge \psi) \to \Box\theta$$

and a similar argument gives

$$\vdash \ \Box(\theta \wedge \psi) \to \Box\psi$$

which gives the required result.

(ii) Set $\psi := \neg\psi$ in (i) and take the contrapositive of the $\leftarrow$ implication.

(iii) Set $\theta := \phi$ and $\psi := \phi$ in (ii).

(iv) This follows from the tautology

$$\neg\theta \to (\theta \to \phi)$$

using (EN).

(v) $\phi \to (\theta \to \phi)$ is a tautology.

(vi) Combine (iv) and (v) using the tautology

$$(\neg\alpha \to \gamma) \to ((\beta \to \gamma) \to ((\alpha \to \beta) \to \gamma)).$$

(vii) By (vi) and (K).

(viii) Set $\theta := \neg\phi$ and $\phi := \neg\theta$ in (vii) and take the contrapositives of the hypothesis and conclusion.

(ix) Since

$$\vdash \ \Box\phi \to (\Diamond\theta \to \Box\phi)$$

we may combine this with (viii) to get

$$\vdash \ \Box\phi \to (\Diamond\theta \to \Diamond\phi)$$

which is easily transformed into the required result.

8.2 Let $\vdash$ be $\vdash_{\mathsf{K4}}$.

(i) By the 4 axiom in contrapostive form we have

$$\vdash \ \Diamond^2 \to \Diamond\phi$$

so that an application of (EN) gives

$$\vdash \ \Box\Diamond^2\phi \to \Box\Diamond\phi.$$

An instance of 8.1(ix) is

$$\vdash \ \Diamond\phi \to (\Box\psi \to \Diamond\psi)$$

so that (EN) and (K) give

$$\vdash \ \Box\Diamond\phi \to (\Box^2\psi \to \Box\Diamond\psi)$$

and hence a use of axiom 4 gives

$$\vdash \Box \Diamond \phi \to (\Box \psi \to \Box \Diamond \psi)$$

which may be rewritten as

$$\vdash \Box \Diamond \phi \wedge \Box \psi \to \Box \Diamond \psi. \tag{B.6}$$

Setting $\psi := \Diamond \phi$ gives

$$\vdash \Box \Diamond \phi \to \Box \Diamond^2 \phi.$$

(ii) Setting $\psi := \Box \Diamond \phi$ in (B.6) gives

$$\vdash \Box \Diamond \phi \wedge \Box^2 \Diamond \phi \to (\Box \Diamond)^2 \phi$$

so that a use of axiom 4 gives

$$\vdash \Box \Diamond \phi \to (\Box \Diamond)^2 \phi.$$

For the converse, an instance of 8.1(ix) gives

$$\vdash_{\mathsf{K}} (\Diamond \phi \wedge \Box \psi) \to \Diamond \psi$$

so that a couple of uses of (N) and axiom K gives

$$\vdash_{\mathsf{K}} (\Box \Diamond \phi \wedge \Box^2 \psi) \to \Box \Diamond \psi.$$

An instance of this is

$$\vdash_{\mathsf{K}} (\Box \Diamond \phi \wedge \Box^3 \Diamond \phi) \to (\Box \Diamond)^2 \phi$$

so that the required result follows by the 4 axiom

8.3 (a) The required inclusions follow from

$$\mathsf{KD} \le \mathsf{KT} \le \mathsf{KDB4}$$

both of which are proved in the chapter.

(b) It is easy to produce:

(i) a one element structure which is symmetric and transitive but not reflexive;

(ii) a two element structure which is reflexive and transitive but not symmetric;

(iii) a three element structure which is reflexive and symmetric but not transitive.

These witness that the inclusions (i,ii,iii) in the diagram

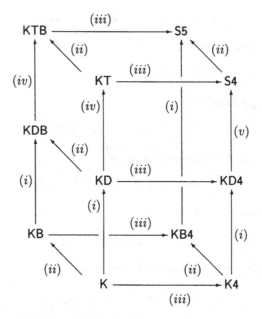

are distinct. For instance, the models of **KB4** are precisely the symmetric, transitive relations and the models of **KT** are precisely the reflexive relations so that **KT** $\not\leq$ **KB4**. Simliarly, **KDB** $\neq$ **KB** otherwise

$$\text{KT} \leq \text{S5} = \text{KDB4} = \text{KB4}$$

etc.

(iv) The structure

$$\circ \longleftrightarrow \circ$$

is serial and symmetric but not reflexive. Hence **KT** $\not\leq$ **KDB** and neither of the two edges (iv) collapse.

(v) The structure

$$\circ \longrightarrow \bullet$$

is serial and transitive but not reflexive. Hence **KT** $\not\leq$ **KD4** and the edge (v) does not collapse.

8.4 By Lemmas 8.11 and 8.12 we have **KB4** = **KB5** so that

$$\text{KB45} = \text{KB4} \quad , \quad \text{KDB5} = \text{KDB4} = \text{S5}$$

and

$$\text{KT5} = \text{KTB5} = \text{KT45} = \text{KTB45} = \text{S5}.$$

This leaves just four new systems, as named, arranged as

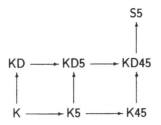

with $\mathsf{K4} \leq \mathsf{K45} \leq \mathsf{KB4}$.

8.5 (a) For each formula ϕ the axiom T and its contrapositive gives

(i) $\Box\phi \to \phi$, $\vdash \phi \to \Diamond\phi$ (ii)

and then

(iii) $\vdash \Box\Diamond\Box\phi \to \Diamond\Box\phi$, $\vdash \Box\Diamond\phi \to \Diamond\Box\Diamond\phi$ (iv)

are particular cases of these. Uses of (EN) on (i,ii) now gives

$\vdash \Diamond\Box\phi \to \Diamond\phi$, $\vdash \Box\phi \to \Box\Diamond\phi$

and then

(v) $\vdash \Box\Diamond\Box\phi \to \Box\Diamond\phi$, $\vdash \Diamond\Box\phi \to \Diamond\Box\Diamond\phi$ (vi)

by a second use of (EN). Two instances of (i,ii) are

(vii) $\vdash \Box\Diamond\phi \to \Diamond\phi$, $\vdash \Box\phi \to \Diamond\Box\phi$ (viii)

so that

(ix) $\vdash \Diamond\Box\Diamond\phi \to \Diamond^2\phi$, $\vdash \Box^2\phi \to \Box\Diamond\Box\phi$ (x)

follow by (EN). But axioms T and 4 give

$\vdash \Box^2 \leftrightarrow \Box\phi$, $\vdash \Diamond^2\phi \leftrightarrow \Diamond\phi$

and so we obtain (xi, xii).

(b) Setting $\phi := \Diamond\phi$ in (xii) and (v) gives

$\vdash \Box\Diamond\phi \to (\Box\Diamond)^2\phi$, $\vdash (\Box\Diamond)^2 \to \Box\Diamond^2\phi$

which since "$\Diamond^2 = \Diamond$" gives the required equivalences.

Now consider any sequence of modal operators

$$\clubsuit \Diamond \heartsuit \spadesuit$$

where each suit is either $\Box$ or $\Diamond$. (The more observant among you will not need to be told that '$\Diamond$' and '$\heartsuit$' are different symbols.) In this sequence, either two consecutive symbols are the same or the symbols alternate. In all cases there is a collapse to a sequence of three or fewer symbols. Thus the diagram exhibits all modal variants of ϕ.

(c) Consider the case where ϕ is a variable P. We must show that there is no further collapse in the diagram. The models of **S4** are precisely the pre-orders i.e. the reflexive transitive relations.

Consider the two element structure

$$a \longrightarrow a \longrightarrow b \longrightarrow b$$

with indicated transition (and no others). Consider also a valuation with $a \Vdash \neg P$ and $b \Vdash P$. Then

$$a \Vdash \neg(\Box \Diamond \Box P \to P).$$

so that

$$\text{not}[\vdash \Box \Diamond \Box P \to P].$$

Thus there is no collapse of (ii) or (xii).

On the same structure consider a valuation with $a \Vdash P$ and $b \Vdash \neg P$. This shows that

$$\text{not}[\vdash P \to \Diamond \Box \Diamond P]$$

so there is no collapse of (i) or (xi).

Adjoin to the above structure a new element c with transitions

$$a \longrightarrow c \longrightarrow c$$

(and no others) and consider a valuation with

$$a \Vdash P \quad , \quad b \Vdash P \quad , \quad c \Vdash \neg P.$$

Then

$$a \Vdash \neg(\Diamond \Box P \to \Box \Diamond P)$$

so that there is no collapse of (iii) or (iv).

Finally consider the relation $\longrightarrow$ on $\mathbb{N}$ where

$$a \longrightarrow b \Leftrightarrow a \geq b$$

(for $a, b \in \mathbb{N}$) and consider a valuation where

$$x \Vdash P \Leftrightarrow x \text{ is even}.$$

Then
$$0 \Vdash \neg(\,\Box \Diamond p \to \Diamond \,\Box P)$$
and there is no collapse of (v) or (vi).

8.6 (a) The axiom 5 gives us

$$\begin{array}{ll}
(l) & \vdash \Diamond \phi \to \Box \Diamond \phi \\
(r) & \vdash \Diamond \Box \phi \to \Box \phi
\end{array}$$

and Exercise 8.1(viii,ix) gives

$$\begin{array}{ll}
(\lambda) & \vdash (\Diamond \theta \to \Box \phi) \to (\Diamond \theta \to \Diamond \phi) \\
(\rho) & \vdash (\Diamond \theta \to \Box \phi) \to (\Box \theta \to \Diamond \phi)
\end{array}$$

We now proceed as follows.

(i)	From (r).	(ii)	By (EN) on (i).
(iii)	By (EN) on (i).	(iv)	An instance of (l).
(v)	An instance of (r).	(vi)	By (iv) and (ii).
(vii)	By (EN) on (vi).	(viii)	By (iv) and (vii).
(ix)	By (iv) and (ρ).	(xii)	By (v) and (λ).
(xiii)	By (viii) and (ρ).	(xiv)	By (viii) and (λ).

(b) Consider the case where ϕ is a variable P. We show that there is no further collapse of the diagram. Recall that the **K5** models are precisely the euclidean relations.

Consider first the three element set $\{a, b, c\}$ structured by the relation

$$a \longrightarrow b \longrightarrow c \longrightarrow c \quad , \quad a \longrightarrow c \longrightarrow b \longrightarrow b$$

together with a valuation such that

$$b \Vdash P \quad , \quad c \Vdash \neg P.$$

Then
$$a \Vdash \Diamond P \quad , \quad a \Vdash \Box P$$
so that
$$\mathrm{not}[\vdash \Diamond p \to \Box P]$$
and hence
$$\mathrm{not}[\vdash \Box \Diamond P \to \Box P] \quad , \quad \mathrm{not}[\vdash \Diamond P \to \Diamond \,\Box P].$$

Next consider the same three element set structured by

$$a \longrightarrow b \longrightarrow b \longrightarrow c \longrightarrow c$$

with a valuation such that

$$b \Vdash P \quad , \quad c \Vdash \neg P.$$

Then

$$a \Vdash \Box P \quad , \quad a \Vdash \neg \Box^2 P$$

so that

$$\text{not}[\vdash \Box P \to \Box^2 P] \quad , \quad \text{not}[\vdash \Diamond^2 P \to \Diamond P].$$

To show that

$$\text{not}[\vdash \Box^2 P \to \Diamond \Box P] \quad , \quad \text{not}[\vdash \Box \Diamond P \to \Diamond^2 P].$$

consider the empty relation on a 1-element set.

Finally, two different valuations on an appropriate 2-element structure shows that

$$\text{not}[\vdash \Diamond \Box P \to P] \quad , \quad \text{not}[\vdash P \to \Box \Diamond P].$$

8.7 Since S5 is S4 extended by the addition of axiom 5, the S5 diagram is a collapsed version of the S4 diagram. The 5 axiom gives

$$\vdash \Diamond \phi \to \Box \Diamond \phi \quad , \quad \vdash \Diamond \Box \phi \to \Box \phi$$

(where $\vdash$ is $\vdash_{S5}$) and these cause the implications (iv),(xi),(iii), and (xii) of the S4 diagram to collapse to equivalences. Thus we obtain the diagram

$$\Box \phi \longrightarrow \phi \longrightarrow \Diamond \phi.$$

These can be shown to be strict using a suitable (small) equivalence relation.

8.8 The first of these follows easily by induction using the T axiom.

For the second consider

$$A = \{0, 1, \ldots, n+1\}$$

structured by the relation $\longrightarrow$ given by

$$x \longrightarrow y \Leftrightarrow \mid x - y \mid \le 1.$$

This relation is reflexive and symmetric and so provides a model of **KTB**. Taking the variable P true at $0, 1, \ldots, n$ and false at $n+1$ gives

$$0 \Vdash \Box^n P, \ \Diamond^{n+1} \neg P$$

so that $\text{not}[\vdash \Box^n P \to \Box^{n+1} P]$, as required.

8.9 Since K5 ≤ K45 and K5 ≤ KB4, in both cases we require a certain collapse of the K5 diagram. The appropriate diagrams are as follows.

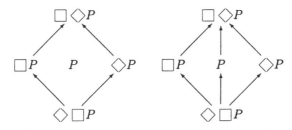

8.10 The temporal structure $(\mathsf{N}, <)$ models none of (ii,iii,iv,vi). The three element temporal structure

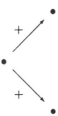

models none of (i,v,vii,viii).

8.11 All of these are confluence properties and only (iii) and (iv) are modelled by all temporal structures.

8.12 (b) (i) ⇒ (ii). Given a wedge

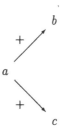

and a variable P, consider a valuation such that

$$x \Vdash P \Leftrightarrow b \sim x$$

(for $x \in A$). Thus $b \Vdash [\approx]P$ and hence, using (i), $b \Vdash [-][+]P$ so that $c \Vdash P$.
(i) ⇒ (iii) Using the same set up observe that $a \Vdash \langle + \rangle [\approx]P$.
The implication (ii) ⇒ (i) and (iii) ⇒ (i) are proved using similar arguments.

(c) By definition, the relation $\approx$ is always reflexive and symmetric. To show transitivity consider any elements a, b, c with

$$c \approx a \approx b.$$

We require $b \approx c$. To show this consider the 9 possible cases. As an example of a non-trivial case, note that if

$$c \xrightarrow{\ -\ } a \xrightarrow{\ +\ } b$$

then the wedge property immediately gives $b \approx c$.

This shows that $\approx$ is an equivalence relation, and, trivially, it includes both $\xrightarrow{+}$ and $\xrightarrow{-}$. Conversely, suppose $\approx$ is any equivalence relation which includes $\xrightarrow{+}$. Then, for each $a, b \in A$,

$$
\begin{aligned}
a \approx b \ &\Rightarrow\ a \xrightarrow{\ -\ } b \text{ or } a = b \text{ or } a \xrightarrow{\ +\ } b \\
&\Rightarrow\ b \xrightarrow{\ +\ } a \text{ or } a = b \text{ or } a \xrightarrow{\ +\ } b \\
&\Rightarrow\ \quad b \sim a \text{ or } a = b \text{ or } a \approx b \quad \Rightarrow\ a \approx b
\end{aligned}
$$

as required.

(d) Immediate.

8.13 (a) $\langle + \rangle \top$

(b) $\theta := [+]\langle - \rangle [+] \bot$, $\psi := \langle - \rangle \top$

(c) R_+

(d) There is no such formula. The easiest way to see this is to consider the associated modal algebras of $\mathcal{Q}$ and $\mathcal{R}$.

8.14 (a) (i)$\Rightarrow$(ii) Suppose that there is some function **next**. Then for each element a and formula ϕ, we have

$$a \Vdash \square \phi \ \Leftrightarrow\ \textbf{next}(a) \Vdash \phi \ \Leftrightarrow\ a \Vdash \Diamond \phi$$

which gives (ii).

(ii)$\Rightarrow$(i) Conversely, for each element a we have $a \Vdash \square \top$, so that (ii) gives $a \Vdash \Diamond \top$, and hence there is at least one b with $a \longrightarrow b$. Suppose there are distinct b and c with $a \longrightarrow b$ and $a \longrightarrow c$. Consider any valuation α such that

$$b \Vdash P \quad,\quad c \Vdash \neg P$$

(for a variable P). Then b witnesses that $a \Vdash \Diamond P$, so that (ii) gives $a \Vdash \square P$, and hence $c \Vdash P$. This is contradictory.

(b) The verification of these shape is mostly routine. For instance, to verify the last shape suppose that

$$a \Vdash [\bullet](\phi \to \bigcirc \phi) \quad,\quad a \Vdash \phi.$$

The first of these gives

$$\text{next}^r(a) \Vdash \phi \Rightarrow \text{next}^{r+1}(a) \Vdash \phi$$

for all $r \in \mathbb{N}$, and hence, using the second, induction gives

$$\text{next}^r(a) \Vdash \phi$$

for all $r \in \mathbb{N}$. This is $a \Vdash [\bullet]\phi$.

8.15 Suppose that $s \in \sigma \cap \tau$. Since $(s, S) \in A$, this gives a wedge

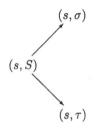

so that, since $\mathcal{A}$ models the confluence shape, we obtain a wedge

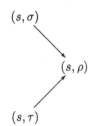

which provides the required ρ with $s \in \rho \subset \sigma \cap \tau$. The proof of the converse is similar.

8.16 (a) For each formula ϕ and compound label i, sufficiently many applications of the (N) rules give

$$\phi \vdash_S [\text{i}]\phi.$$

The first and last implications hold since $\Phi \subseteq \Phi^*$. For the central implication, if $\Phi^* \vdash_{\overline{S}}^w \phi$ then $\vdash_S [1]\theta_1 \wedge \cdots \wedge [n]\theta_n \rightarrow \phi$ for some $\theta_1, \ldots, \theta_n \in \Phi$ and composite labels i(1),..., i(n). (For convenience, only the index of the compound label has been shown in the corresponding box.) But then, by the first part, we have

$$\Phi \vdash_S [1]\theta_1 \wedge \cdots \wedge [n]\theta_n$$

so that $\Phi \vdash_S \phi$ follows by MP.

For each θ we have $\square\theta \in \{\theta\}^*$ so that

$$\{\theta\}^* \vdash_{\overline{S}}^w \square\theta.$$

However, $\theta \vdash_S^w \; \Box \theta$ holds (in general) only if S is pathetic. Thus the left hand implication is not reversible.

(b) The implication

$$\Phi^* \vdash_S \phi \; \Rightarrow \; \Phi^* \vdash_S^w \phi$$

is proved by induction on the size of the hypothesis set Φ. The induction step relies on the fact that if

$$\Phi^* \cup \{\theta\}^* \vdash_S \phi \qquad\qquad (B.7)$$

then there is some compound label i with

$$\Phi^* \vdash_S \; [i] \to \phi.$$

This is proved by induction on the length of the given witnessing deduction. The crucial induction step is the one across the use of the (N) rule.

To verify this step suppose that

$$\phi = [j]\psi$$

for some label j and formula ψ, suppose that (B.7) is obtained from

$$\Phi^* \cup \{\theta\}^* \vdash_S \psi$$

by an application of rule (Nj). The Induction Hypothesis gives us some compound label i with

$$\Phi^* \vdash_S \; [i]\theta \to \psi$$

so that (ENj) gives

$$\Phi^* \vdash_S \; [j] \, [i]\theta \to \phi$$

as required.

(c) This holds since

$$\Psi \vDash_S^k \phi \; \Leftrightarrow \; \Psi \cup \mathcal{S} \vDash \phi$$

and $\mathcal{S}^* = \mathcal{S}$ (where $\mathcal{S}$ is the set of axioms of S).

B.9   Chapter 9

9.1 For instance since s is closed under implication we have

$$\left.\begin{array}{r} \theta \to \psi \in s \\[2mm] \theta \in s \end{array}\right\} \; \Rightarrow \; \psi \in s$$

i.e.

$$\theta \to \psi \in s \;\Rightarrow\; \theta \notin s \text{ or } \psi \in s.$$

Conversely, if $(\theta \to \psi) \notin s$ then there is some conjunction σ of finitely many members of s with

$$\vdash_S \sigma \to \neg(\theta \to \psi)$$

i.e.

$$\vdash_S \sigma \to \theta \wedge \neg\psi$$

so that $\theta \in s$ and $\neg\psi \in s$.

9.2 (a) Only one implication is non-trivial. For a given $s \in \mathbf{S}(\Phi)$, label i, and formula ϕ, suppose that

$$s \xrightarrow{\;i\;} t \;\Rightarrow\; \phi \in t$$

holds for all $t \in \mathbf{S}(\Phi)$. Let Ψ be the set of all formulas given by

$$\psi \in \Psi \;\Leftrightarrow\; [i]\psi \in s.$$

Then, for each $t \in \mathbf{S}$

$$\Phi^* \cup \Psi \subseteq t \quad\Rightarrow\quad t \in \mathbf{S}(\Phi) \text{ and } s \xrightarrow{\;i\;} t \quad\Rightarrow\quad \phi \in t$$

so that Lemma 9.5 gives

$$\Phi^* \cup \Psi \vdash_S^w \phi.$$

Thus there are $\theta_1, \ldots, \theta_m \in \Phi^*$ and $\psi_1, \ldots, \psi_n \in \Psi$ with

$$\vdash_S \theta_1 \wedge \cdots \theta_m \wedge \psi_1 \wedge \cdots \wedge \psi_n \to \phi$$

so that

$$\vdash_S [i]\theta_1 \wedge \cdots [i]\theta_m \wedge [i]\psi_1 \wedge \cdots \wedge [i]\psi_n \to [i]\phi.$$

But for each of these θ we have

$$[i]\theta \in \Phi^* \subseteq s$$

so that

$$[i]\theta_1 \wedge \cdots \wedge [i]\theta_m \in s$$

and a similar argument shows that

$$[i]\psi_1 \wedge \cdots \wedge [i]\psi_n \in s$$

and hence $[i]\phi \in s$.

(b) This is proved by induction on the complexity of ϕ. For the induction step across $[i]$ we argue

$$s \Vdash [i]\phi \;\Leftrightarrow\; \text{For each } t \in \mathbf{S}(\Phi),$$
$$s \xrightarrow{\;i\;} t \;\Rightarrow\; t \Vdash \phi$$

$$\Leftrightarrow\; \text{For each } t \in \mathbf{S}(\Phi),$$
$$s \xrightarrow{\;i\;} t \;\Rightarrow\; \phi \in t \quad\Leftrightarrow\quad [i]\phi \in s$$

where the second equivalence follows by the Induction Hypothesis and the third by (a).

(c) Since $S \cup \Phi^* \subseteq s$ for all $s \in S(\Phi)$.

(d) The two implications

$$(i) \Rightarrow (ii\ l) \Rightarrow (iii\ l)$$

hold by Exercise 8.16(a). The two implications

$$(ii\ l) \Rightarrow (ii\ r) \quad , \quad (iii\ l) \Rightarrow (iii\ r)$$

are standard soundness results, and the implication (ii r)$\Rightarrow$(iii r) is trivial (since $\Phi \subseteq \Phi^*$). The implication (iii r)$\Rightarrow$(iv) holds by (c).

Finally, to prove the implication (iv)$\Rightarrow$(i), if $(\mathfrak{S}(\Phi), \sigma)$ models ϕ then ϕ is in every member of $S(\Phi)$ so that Lemma 9.5 gives (i).

B.10   Chapter 10

10.1 If $(\mathfrak{S}, \sigma)$ models any sentence, then $\mathfrak{S}$ itself also models that sentence.

10.2 For the shape E the corresponding structural property is that

$$a \longrightarrow b \Rightarrow a \longrightarrow a$$

holds for all elements a and b. To see that the canonical structure $\mathfrak{S}$ has this property consider any $s, t \in S$ with

$$s \longrightarrow t.$$

Since $\top \in T$, this gives $\Diamond \top \in s$. Hence, for each formula ϕ,

$$\Box \phi \in s \Rightarrow \Diamond \top \wedge \Box \phi \in s \Rightarrow \phi \in s$$

as required.

For the shape F the corresponding structural property is that

$$a \longrightarrow b \Rightarrow b \longrightarrow b$$

holds for all elements a and b. To see that the canonical structure $\mathfrak{S}$ has this property consider any $s, t \in S$ with

$$s \longrightarrow t.$$

Since $\Box(\Box \phi \to \phi) \in s$, this means that $(\Box \phi \to \phi) \in t$ and hence

$$\Box \phi \in t \Rightarrow \phi \in t$$

so that (since ϕ is arbitrary) $t \longrightarrow t$, as required.

10.3 Let $\mathsf{S} = \mathsf{K}(i, j, k, l, m, n)$. We refer to the corresponding structural property as given in Exercise 5.3. Thus consider any $r, s, t \in \boldsymbol{S}$ with

$$r \xrightarrow{\ l\ } s \xrightarrow{\ n\ } t$$

and set

$$\Xi := \{\theta \mid [i]\theta \in r\} \cup \{[j]\phi \mid [m]\phi \in s\} \cup \{\langle k \rangle \psi \mid \psi \in t\}.$$

We show that Ξ is S-consistent.

If Ξ is not S-consistent then there are finite sets Θ, Φ, Ψ of formulas taken from the appropriate places where

$$\vdash_{\mathsf{S}} \bigwedge \Theta \wedge \bigwedge [j]\Phi \wedge \bigwedge \langle k \rangle \Psi \ \rightarrow \ \bot.$$

Since box operators commute with $\wedge$ we may reduce Θ and Φ to single formulas θ and ϕ to get

$$\vdash_{\mathsf{S}} \theta \rightarrow ([j]\phi \rightarrow \bigvee [k]\neg\Psi)$$

which gives

$$\vdash_{\mathsf{S}} \theta \rightarrow ([j]\phi \rightarrow [k] \bigvee \neg\Psi)$$

and hence

$$\vdash_{\mathsf{S}} [i]\theta \rightarrow [i]([j]\phi \rightarrow [k] \bigvee \neg\Psi).$$

Since, by choice, we have $[i]\theta \in r$, this gives

$$[i]([j]\phi \rightarrow [k] \bigvee \neg\Psi) \in r$$

hence, using the axiom,

$$[l]([m]\phi \rightarrow [n] \bigvee \neg\Psi) \in r.$$

From this we have

$$([m]\phi \rightarrow [n] \bigvee \neg\Psi) \in s$$

so that, by choice of ϕ, we have

$$[n] \bigvee \neg\Psi \in s$$

and hence

$$\bigvee \neg\Psi \in t.$$

This produces some $\psi \in \Psi \subseteq t$ with $\neg\psi \in t$, which is the required contradiction.

The consistency of Ξ gives us some $u \in \boldsymbol{S}$ with $\Xi \subseteq u$, and hence

$$[i]\theta \in r \Rightarrow \theta \in u \quad , \quad [m]\phi \in s \Rightarrow [j]\phi \in u \quad , \quad \psi \in t \Rightarrow \langle k \rangle \psi \in u$$

(for all θ, ϕ, ψ). The first and third of these ensure that

$$r \xrightarrow{\ i\ } u \xrightarrow{\ k\ } t.$$

Also, for any $v \in \boldsymbol{S}$ with

$$u \xrightarrow{\ j\ } v$$

the second gives

$$[m]\phi \in s \ \Rightarrow \ [j]\phi \in u \ \Rightarrow \ \phi \in v$$

so that

$$s \xrightarrow{\ m\ } v$$

as required.

10.4 Consider any $r, s, t \in \boldsymbol{S}$ with

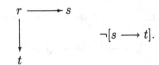

We must show that $t \longrightarrow s$.

Note first that there is a formula ψ with

$$\Box\psi \in s \ \ , \ \ \neg\psi \in t.$$

Now consider the set of formulas

$$\Phi \ = \ s \cup \{\phi \mid \Box\phi \in t\}.$$

If this is not S-consistent, then there are

$$\theta \in s \ \ , \ \ \Box\phi \in t$$

with

$$\vdash_{\mathsf{S}} \theta \wedge \phi \to \bot.$$

But then $\Box\phi, \neg\psi \in t$ so that

$$\Box(\Box\phi \to \psi) \notin r$$

and hence, using the axiom

$$\Box\psi \to \phi \in s$$

which gives $\phi \in s$, and (since $\theta \in s$) this leads to a contradiction.

Since Φ is consistent, the maximality of s gives $\Phi = s$, and hence $t \longrightarrow s$.

10.5 This shape corresponds to the structural property of Exercise 6.2(d) of Chapter 6. We must show that the canonical structure $\mathfrak{S}$ has this property. Thus consider $r, s, t, u \in \boldsymbol{S}$ with

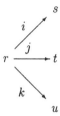

and let

$$\Xi := \{\phi \mid [\imath]\phi \in s\} \cup \{\theta \mid [m]\theta \in t\} \cup \{\psi \mid [n]\psi \in u\}.$$

It suffices to show that Ξ is S-consistent.

If Ξ is not S-consistent then (remembering that box operators commute with $\wedge$) there are formulas ϕ, θ, ψ with

$$\vdash_{\mathsf{S}} \phi \wedge \theta \wedge \psi \to \bot$$

and

$$[\imath]\phi \in s \quad , \quad [m]\theta \in t \quad , \quad [n]\psi \in u.$$

The assumed configuration gives

$$\langle\imath\rangle[\imath]\phi \in r \quad , \quad \langle k\rangle[n]\psi \in r$$

so that, using the axiom, we have

$$[j]\langle m\rangle(\phi \wedge \psi) \in r$$

and hence

$$\langle m\rangle(\phi \wedge \psi) \in t.$$

But

$$\vdash_{\mathsf{S}} \phi \wedge \psi \to \neg\theta$$

so that

$$\vdash_{\mathsf{S}} \langle m\rangle(\phi \wedge \psi) \to \langle m\rangle\neg\theta$$

and hence $\langle m\rangle\neg\theta \in t$. This contradicts the original choice of θ.

10.6 We use the correspondence result given in Exercise 6.3. Thus we must show that the canonical structure $\mathfrak{S}$ has this property.

To this end consider points $a, b, c(1), c(2), \ldots, d$ of $\mathfrak{S}$ with

$$a \xrightarrow{\ i\ } b \xrightarrow{\ l\ } d \quad \text{and} \quad b \xrightarrow{\ j(p)\ } c(p) \quad \text{for } p = 1, 2, \ldots$$

(where, of course, p varies over a finite set). For each $p = 1, 2, \ldots$ let

$$\Phi(p) = \{\phi \mid [k]\phi \in c(p)\} \quad \text{(where } k \text{ is } k(p)\text{)}$$

and let

$$\Psi = \{\psi \mid [m]\psi \in d\}.$$

These sets are closed under conjunction.

It suffices to show that

$$\Psi \cup \{\langle n \rangle \top\} \cup \Phi(i) \cup \Phi(2) \cup \cdots$$

is consistent. (The consistency of

$$\Psi \cup \Phi(1) \cup \Phi(2) \cup \cdots$$

ensures the existence of the required point e, and the extra component $\langle n \rangle \top$ ensures the existence of the required f).

If this set is not consistent then there are $\psi \in \Psi, \phi_1 \in \Phi(1), \phi_2 \in \Phi(2), \ldots$ with

$$\vdash_S \psi \wedge \langle n \rangle \top \wedge \bigwedge \Phi \rightarrow \bot$$

where

$$\Phi = \{\phi_p \mid p = 1, 2, \ldots\}.$$

This gives

$$\vdash_S \langle n \rangle \top \wedge \bigwedge \Phi \rightarrow \neg\psi$$

and hence

$$\vdash_S \langle m \rangle (\langle n \rangle \top \wedge \bigwedge \Phi) \rightarrow \langle m \rangle \neg\psi$$

which, by the choice of ψ, gives

$$\neg \langle m \rangle (\langle n \rangle \top \wedge \bigwedge \Phi) \in d.$$

We now produce a contradiction of this.

For each $p = 1, 2, \ldots$ we have

$$[k]\phi \in c(p) \quad \text{(where } k \text{ is } k(p)\text{)}$$

so that

$$\langle p \rangle [p]\phi_p \in b$$

and hence

$$\bigwedge \{\langle p \rangle [p]\phi_p \mid p = 1, 2, \ldots\} \in b.$$

Since $a \xrightarrow{i} b$ the assumed axiom gives

$$[i]\langle m \rangle (\langle n \rangle \top \wedge \bigwedge \Phi) \in b$$

and hence

$$\langle m \rangle (\langle n \rangle \top \wedge \bigwedge \Phi) \in d$$

which is the required contradiction.

10.7 (a) The relation $\longrightarrow$ of $\mathcal{N}$ is transitive and well founded, so $\mathcal{N}$ models $L_-(\phi)$. To show that $\mathcal{N}$ models $L_+(\phi)$ consider any valuation on $\mathcal{N}$, formula ϕ and $m \in \mathbf{N}$ with

$$m \Vdash \langle + \rangle [+] \phi \quad , \quad m \Vdash [+] ([+] \phi \to \phi).$$

We require that $m \Vdash [+] \phi$.
We know there is some $n \in \mathbf{N}$ with

$$m + n \Vdash [+] \phi.$$

Take the least such n. We show that this $n = 0$. To do this note that if $n \neq 0$ then we also have

$$m + n \Vdash [+] \phi \to \phi$$

so that $m + n \Vdash \phi$. Thus, with $n = k + 1$ we have

$$m + k \Vdash [+] \phi$$

which contradicts the minimality of n.

(b) (i) By way of contradiction suppose there is some $b \in A$ such that for each $x \in A$

$$b \sim x \ \Rightarrow \ (\exists y)[y \overset{+}{\longrightarrow} x].$$

This enables us to produce an infinite chain

$$b = b_0 \overset{-}{\longrightarrow} b_1 \overset{-}{\longrightarrow} b_2 \overset{-}{\longrightarrow} \cdots \tag{B.8}$$

which contradicts the characteristic property associated with L_-.
For uniqueness note that if $a_1, a_2 \in I(\mathcal{A})$ satisfy $a_1 \sim a_2$, then

$$a_2 \overset{-}{\longrightarrow} a_1 \ \text{or} \ a_1 = a_2 \ \text{or} \ a_1 \overset{-}{\longrightarrow} a_2$$

and the first and last of these are prevented by the defining property of $I(\mathcal{A})$.

(ii) By way of contradiction, if there is no such function, then there is some $a \in A$ such that

$$a \overset{+}{\longrightarrow} x \ \Rightarrow \ (\exists y)[a \overset{+}{\longrightarrow} y \overset{+}{\longrightarrow} x].$$

Taking any element b with $a \overset{+}{\longrightarrow} b$ this enables us to produce an infinite chain (B.8) where

$$a \overset{+}{\longrightarrow} b_{r+1} \overset{+}{\longrightarrow} b_r$$

for all $r < \omega$. This contradicts L_-.

(iii) Consider $a, b \in A$ with

$$\mathsf{next}(a) = \mathsf{next}(b) = c \quad (\text{say}).$$

Then $a \sim c \sim b$ so that $a \sim b$ and hence either $a = b$ (which is what we want) or

$$a \xrightarrow{+} b \quad , \quad \text{so that } b = \mathsf{next}(a) \text{ or } \mathsf{next}(a) \xrightarrow{+} b$$

or

$$b \xrightarrow{+} a \quad , \quad \text{so that } a = \mathsf{next}(b) \text{ or } \mathsf{next}(b) \xrightarrow{+} a.$$

In both the latter two cases we find that $c \xrightarrow{+} c$ which is a contradiction since, by L_- both the relations $\xrightarrow{-}$ and $\xrightarrow{+}$ are irreflexive.

(c) (i) Note first that for each $x \in A$,

$$x \Vdash P \;\Rightarrow\; x \Vdash \langle + \rangle P$$

(for if $m \in \mathbb{N}$ witnesses the left hand side then $m + 1$ witnesses the right hand side). Thus $x \Vdash [+]\neg P \to \neg P$, and in particular

$$a \Vdash [+]([+]\neg P \to \neg P).$$

The axiom base on L_+ now gives

$$\Vdash \langle + \rangle [+]\neg P \to [+]\neg P$$

which, since $a \Vdash \langle + \rangle P$, gives the required result.

(ii) For each $b \in A$ there is a unique $a \in A$ with $a \sim b$, But then either $a = b$ (and we may take $k = 0$) or $a \xrightarrow{+} b$. By (ii) the second of these gives the required $m \in \mathbb{N}$. Uniqueness follows by the injectivity of next.

B.11   Chapter 11

11.1 (a) Easy.

(b) Suppose that

$$\mathcal{A} \xrightarrow{\; f \;} \mathcal{B} \xrightarrow{\; g \;} \mathcal{C}$$

is a pair of p-morphisms, and consider any label i, elements $a \in A, z \in C$ with

$$g(f(a)) \xrightarrow{\; i \;} z.$$

Since g is a p-morphism, this produces some $y \in B$ with

$$f(a) \xrightarrow{\; i \;} y \quad , \quad g(y) = z$$

and then, since f is a p-morphism, there is some $x \in A$ with

$$a \xrightarrow{i} x \quad , \quad f(x) = y.$$

Since $g(f(x)) = g(y) = z$, this shows that gf is a p-morphism.
The proof of the extra condition for zigzag morphisms is similar.

11.2 (a) Since the relation on $\mathcal{R}$ is full, for each $a, b \in A$ we have $g(a) \longrightarrow g(b)$, so that g is a morphism.

Suppose that $\mathcal{A}$ is serial and consider any $a \in A, y \in \{0\}$ with $g(a) \longrightarrow y$. Seriality gives us some $x \in A$ with $a \longrightarrow x$, and then $g(x) = 0 = y$, so that g is a p-morphism.

Conversely, suppose that g is p-morphism and consider any $a \in A$. Since $g(a) \longrightarrow 0$, there is some $x \in A$ with $a \longrightarrow x$ and $g(x) = 0$. Thus $\mathcal{A}$ is serial.

(b) For each $a \in A$ let f_a be the assignment $0 \mapsto a$. These are all the functions $\{0\} \longrightarrow A$, and since $\neg[0 \longrightarrow 0]$ in $\mathcal{L}$, each one is a morphism. The assignment f_a is a p-morphism precisely when for each $y \in A$ with $a = f_a(0) \longrightarrow y$, there is some $x \in \{0\}$ with $0 \longrightarrow x$ and $f_a(x) = y$, i.e. when $0 \longrightarrow 0$ and $y = f_a(0) = a$. Thus f_a is a p-morphism precisely when there is no $y \in A$ with $a \longrightarrow y$, i.e. when a is blind.

(c) The unique assignment $\longrightarrow \{0\}$ is a morphism precisely when the relation of $\mathcal{A}$ is empty. Such a morphism is always a p-morphism.
The assignment $f_a : \{0\} \longrightarrow A$ provides a morphism $\mathcal{R} \longrightarrow \mathcal{A}$ precisely when the element a of $\mathcal{A}$ is reflexive. Such a morphism is a p-morphism precisely when a is isolated, i.e. when

$$a \longrightarrow x \;\Rightarrow\; x = a$$

holds for all $x \in A$.

11.3 These two valuations are given by

$$b \in \lambda(P) \;\Leftrightarrow\; (\exists a \in \alpha(P))[f(a) = b] \quad , \quad b \in \rho(P) \;\Leftrightarrow\; (\forall a \in \alpha(P))[f(a) = b]$$

for all $b \in B$ and variables P.

11.4 (b) Suppose first that $\mathcal{A} \subseteq_g \mathcal{B}$ and consider any $a \in A$ and $y \in B$ with $f(a) \xrightarrow{i} y$, i.e. such that $a \xrightarrow{i} y$ holds in $\mathcal{B}$. Then, in fact, we have $y \in A$, so we may set $x = y$ to verify that f is a p-morphism.

Conversely, suppose that f is a p-morphism, and consider any $a \in A, b \in B$ with $a \xrightarrow{i} b$. Then $f(a) \xrightarrow{i} b$ so there is some $x \in A$ with $a \xrightarrow{i} x$ and $b = f(x) = x$. Thus $b \in A$, and hence $\mathcal{A} \subseteq_g \mathcal{B}$.

11.5 (a) Suppose that both R and S are back-and-forth relations, and consider any $a \in A$ and $c \in C$ with $aR;Sc$. This gives some $b \in B$ with

$$aRbSc.$$

Consider any $z \in C$ with $c \xrightarrow{i} z$. Then, using S, there is some $y \in B$ with

$$b \xrightarrow{i} y \quad , \quad ySz$$

and then, using R, there is some $x \in A$ with

$$a \xrightarrow{i} x \quad , \quad xRy$$

Since

$$xRySz$$

this shows that $R : S$ has the back property, and a similar argument shows that it has the forth property.

The required extension to bisimulations is easy.

(b) For $a \in A$ and $c \in C$ we have

$$aF;Gc \;\Leftrightarrow\; (\exists b \in B)[aFGbGc]$$
$$\Leftrightarrow\; (\exists b \in B)[f(a) = b \text{ and } g(b) = c] \;\Leftrightarrow\; g(f(a)) = c$$

as required.

11.6 (a) We know that $a \sim_0 b$ holds for all $a, b \in \mathsf{N}$. Thus

$$a \sim_1 b$$

holds precisely when for each

$$x \in \mathsf{N} \qquad \text{or} \qquad y \in \mathsf{N}$$

with

$$a = x + 1 \qquad \text{or} \qquad b = y + 1$$

there is some

$$y \in \mathsf{N} \qquad \text{or} \qquad x \in \mathsf{N}$$

with

$$b = y + 1 \qquad \text{or} \qquad a = x + 1$$

and

$$x \sim_0 y.$$

In other words, $a \sim_0 b$ holds if either $a = b = 0$ or both a, b are non-zero. This proves the base case $r = 0$ of the required induction.

To prove the induction step $r \mapsto r + 1$ suppose first that

$$a \sim_{r+2} b.$$

Then, by the definition of $\sim_{r+2}$ as $(\sim_{r+1})^{\blacktriangledown}$, we have

$$a \sim_{r+1} b$$

together with a certain back-and-forth property. Suppose also that $a \leq r + 1$ (so that we require $a = b$). If $a \leq r$ then the Induction Hypothesis gives $a = b$ as required. If $a = r + 1$ then the forth property gives some y with

$$r \sim_{r+1} y \quad \text{and} \quad b = y + 1$$

so that $b = r + 1 = a$ as required.

The required converse is proved in a similar manner.

(b) An easy exercise shows that

$$a \Vdash \Diamond^{k+1} \bot \Leftrightarrow a \leq k$$

so that the sentence $\phi_k := \Diamond^{k+1} \bot \wedge \Box^k \top$ will do.

B.12 Chapter 12

12.1 The left-most filtration of this structure maps it onto a 4 element structure as follows.

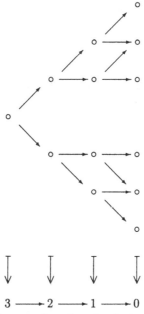

To see this first rank the elements into four levels $3, 2, 1, 0$ as shown. Observe that for each $a, b \in A$,

$$a, b \text{ have the same rank} \Leftrightarrow \begin{array}{c} \text{the substructures generated by} \\ a \text{ and } b \text{ are isomorphic} \end{array}$$

and hence

$$a, b \text{ have the same rank } \Rightarrow a \sim b.$$

Note also that

$$a \text{ has rank } 0 \iff a \Vdash \Box \bot$$
$$a \text{ has rank } 1 \iff a \Vdash \Diamond \Box \bot$$
$$a \text{ has rank } 2 \iff a \Vdash \Diamond^2 \Box \bot$$
$$a \text{ has rank } 3 \iff a \Vdash \Diamond^3 \Box \bot$$

and hence

$$a, b \text{ have the same rank } \iff a \sim b.$$

The required result is now straight forward.

To show that this is also the right-most filtration, consider any $0 \leq m, n \leq 3$ and suppose that for each a of rank m and each b of rank n we have

$$a \Vdash \Box \sigma \Rightarrow b \Vdash \sigma.$$

for all sentences σ. We require $m = n + 1$.

But $m \neq 0$, otherwise there is some a of rank 0 with $a \Vdash \Box \bot$. Let $m = k + 1$ and consider the sentence σ where

$$x \text{ has rank } k \iff x \Vdash \sigma$$

holds for all $x \in A$. Then $a \Vdash \Box \sigma$ for some a of rank m, so that $b \Vdash \sigma$ of rank n, i.e. $n = k$.

12.2 (a) For each $m, n \in \mathbb{N}$, the two substructures of $\mathcal{N}^+$ generated by m and n are isomorphic, hence $m \sim n$.

(b) For each $k \in \mathbb{N}$, there is a sentence σ such that

$$m \Vdash \sigma \iff m = k$$

holds for all elements m of $\mathcal{N}^-$.

12.3 Divide the nodes into four types as follows.

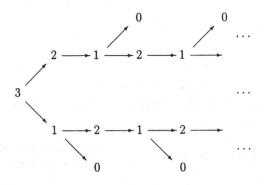

Note that nodes a and b have the same type if and only if their generated substructures are isomorphic. Thus

$$a, b \text{ have the same type } \Rightarrow a \sim b.$$

Observe also that for each node a,

a has rank 0 $\Leftrightarrow$ $a \Vdash \Box\bot$
a has rank 1 $\Leftrightarrow$ $a \Vdash \Diamond\Box\bot$
a has rank 2 $\Leftrightarrow$ $a \Vdash \Box\Diamond\Box\bot$
a has rank 3 $\Leftrightarrow$ $a \Vdash (\Diamond^2\Box\bot \wedge (\Diamond\Box)^2\bot)$.

Thus
$$a, b \text{ have the same type } \Leftrightarrow a \sim b$$

and we see that

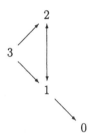

is the leftmost filtration.

12.4 Each isomorphism is a filtration.

12.5 For distinct $a, b \in A$ there is some $P \in \Gamma$ with $a \Vdash P$ and $b \Vdash \neg P$. Thus the induced equivalence relation $\sim$ is just equality, and hence the leftmost Γ-filtration is the identity map

$$(\mathcal{A}, \alpha) \longrightarrow (\mathcal{A}, \alpha).$$

For $b, y \in A$, the right-most transition

$$b \longrightarrow_r y$$

holds precisely when for each formula ϕ with $\Box\phi \in \Gamma, \dots$. Since there are no such ϕ, this condition is vacuously satisfied, and hence the right-most filtration is the complete graph on A.

12.6 (a) Suppose $a \neq x$. Then, since $(\mathcal{A}, \alpha)$ is separated, we have $a \not\approx x$, which gives the required formula $\xi_{a,x}$.

(b) Since A is finite we may set

$$\rho_a = \bigwedge\{\xi_{a,x} \mid x \in A - \{a\}\}$$

to obtain the required ρ_a.

(c) $\tau_X = \bigvee\{\rho_a \mid a \in X\}$

(d) The case when $\phi := P$ follows by (c), and then the result for a general ϕ follows by a routine induction.

12.7 Both zigzag morphism and Γ-filtrations satisfy (Val$^{\leftrightarrow}$), so it suffices to connect (Rel$^{\leftarrow}$) with (Fil).

(a) Consider any $a, x \in A$ with

$$f(a) \xrightarrow{\ i\ } f(x)$$

and any formula ϕ with $a \Vdash [i]\phi$. We use (Rel$^{\leftarrow}$) to show that $x \Vdash \phi$. Thus (Rel$^{\leftarrow}$) gives us some $u \in A$ with

$$a \xrightarrow{\ i\ } u \quad , \quad f(u) = f(x).$$

The zigzag preservation property gives

$$u \Vdash \phi \Leftrightarrow f(u) \Vdash \phi \Leftrightarrow f(x) \Vdash \phi \Leftrightarrow x \Vdash \phi$$

so that

$$a \Vdash [i]\phi \Rightarrow u \Vdash \phi \Rightarrow x \Vdash \phi$$

as required.

(b) Consider any $a \in A$ and $b \in B$ with

$$f(a) \xrightarrow{\ i\ } b.$$

Since B is finite and separated, there is some formula ρ_b such that

$$y \Vdash \rho_b \Leftrightarrow y = b$$

holds for all $y \in B$. In particular, $f(a) \Vdash \langle i \rangle \rho_b$, so that the filtration preservation property gives $a \Vdash \langle i \rangle \rho_b$, which produces some $x \in A$ with

$$a \xrightarrow{\ i\ } x \quad , \quad x \Vdash \rho_b.$$

The filtration preservation property also gives $f(x) \Vdash \rho_b$, and hence $f(x) = b$, which verifies (Rel$^{\leftarrow}$).

B.13   Chapter 13

13.1 Let σ be the sentence which axiomatizes S, and let M be the class of finite models of S. Let ϕ be a formula which is modelled by every member of M. We must show that $\vdash_S \phi$.

Consider any (valued) model $(\mathcal{A}, \alpha)$ of S. It suffices to show that $(\mathcal{A}, \alpha) \Vdash^v \phi$.

To this end let $\Gamma = \Gamma(\phi \wedge \sigma)$ be the set of subformulas of $\phi \wedge \sigma$. Let

$$(\mathcal{A}, \alpha) \xrightarrow{\ f\ } (\mathcal{B}, \beta)$$

be a Γ-filtration (so that $\mathcal{B}$ is finite). Since $(\mathcal{A}, \alpha) \Vdash^v \sigma$, we have $(\mathcal{B}, \beta) \Vdash^v \sigma$, so that (since σ is variable-free) $\mathcal{B} \Vdash^u \sigma$, and hence $\mathcal{B} \in M$. But then, by hypothesis, $\mathcal{B} \Vdash^u \phi$, so that $(\mathcal{B}, \beta) \Vdash^v \phi$, as required.

13.2 Given a transition $b \longrightarrow y$ in the target, there are $a \in b$ and $x \in y$ with $a \longrightarrow x$. But then $a = x$ so that $b = y$.

13.3 The composition property is that

$$a \xrightarrow{\ i\ } b \xrightarrow{\ j\ } c \Rightarrow a \xrightarrow{\ k\ } c$$

(for all elements a, b, c). In particular the given 4-element structure $\mathcal{A}$ vacuously has this property.

Let $\Gamma = \{P, Q\}$ where P, Q are distinct variables. Consider the valuation α on $\mathcal{A}$ indicated by

$$\begin{array}{cccc} \circ \xrightarrow{\ i\ } \circ & & \circ \xrightarrow{\ j\ } \circ \\ P \quad\quad P & & P \quad\quad \neg P \\ \neg Q \quad\quad Q & & Q \quad\quad Q. \end{array}$$

The Γ induced equivalence on $(\mathcal{A}, \alpha)$ coalesces the two central nodes. Thus

$$\circ \xrightarrow{\ i\ } \circ \xrightarrow{\ j\ } \circ$$

gives the left-most filtration. This structure does not have the composition property.

13.4 (a)(i) This shape captures the property that: For each element a, there is some b with $a \xrightarrow{\ i\ } b$ and $b \xrightarrow{\ j\ } b$. This property is preserved by all surjective morphisms.

(ii) This captures the property that: For all a and b,

$$a \xrightarrow{\ j\ } b \Rightarrow a \xrightarrow{\ i\ } b.$$

Consider any elements x, y of the source of the filtration with

$$f(x) \xrightarrow{\;j\;} f(y)$$

(where f is the filtration morphism). Then there are $a \in f(x), b \in f(y)$ with $a \xrightarrow{\;j\;} b$, the assumed property gives $a \xrightarrow{\;i\;} b$ hence, by the morphism property

$$f(x) = f(a) \xrightarrow{\;i\;} f(b) = f(y)$$

as required.

(iii) This captures the property that: For all a and b,

$$a \xrightarrow{\;j\;} b \;\Rightarrow\; b \xrightarrow{\;i\;} a$$

(for all elements a and b). An argument similar to the one for (ii) shows the required preservation.

(iv) This captures the property that: For each configuration

$$a \xrightarrow{\;j\;} b$$

there is some element c with

$$a \xrightarrow{\;i\;} b \xrightarrow{\;k\;} c.$$

Again a similar argument to the one for (ii) gives the required preservation.

(b) To use this argument we need to consider models of both

$$[i]\phi \to [j]\phi \quad \text{and} \quad [j]\phi \to [i][i]\phi.$$

But this second shape is **not** preserved by left-most filtrations.

13.5 (a) We know that a structure $\mathcal{A}$ models **KE** precisely when

$$a \longrightarrow b \;\Rightarrow\; a \text{ is reflexive}$$

holds for all $a, b \in A$. We show that this property is preserved by left-most filtrations.

Thus, consider any such filtration

$$(\mathcal{A}, \alpha) \xrightarrow{\;f\;} (\mathcal{B}, \beta)$$

where $\mathcal{A}$ models **KE**. Consider any $b_1, b_2 \in B$ with

$$b_1 \longrightarrow b_2.$$

Then there are $a_1 \in b_1$ and $a_2 \in b_2$ with

$$a_1 \longrightarrow a_2$$

and hence $a_1 \longrightarrow a_1$ so that the morphism property gives $b_1 \longrightarrow b_1$, as required.

(b) These two results follow by Lemmas 13.7 and 13.8. Note, however, that KDE = KT.

(c) For a model $\mathcal{A}$ of KE4 and a valuation α on $\mathcal{A}$, consider the usual set up

$$A \xrightarrow{\ f\ } B$$

induced by a set Γ of formulas. Let $\longrightarrow$ be the transition relation on B defined by

$$b_1 \longrightarrow b_2 \Leftrightarrow \begin{cases} \text{For all } a_1 \in b_1, a_2 \in b_2 \text{ and} \\ \text{formulas } \phi \text{ with } \Box\phi \in \Gamma, \\ a_1 \Vdash \Box\phi \Rightarrow a_1 \Vdash \phi \\ \text{and} \\ a_1 \Vdash \Box\phi \Rightarrow a_2 \Vdash \phi \wedge \Box\phi. \end{cases}$$

Consider the usual valuation β on this constructed transition structure $\mathcal{B}$. We show first that f provides a morphism

$$(\mathcal{A}, \alpha) \xrightarrow{\ f\ } (\mathcal{B}, \beta).$$

Thus, consider any $x_1, x_2 \in A$ with

$$x_1 \longrightarrow x_2$$

(so that x_1 is automatically reflexive) and any $a_1 \in f(x_1), a_2 \in f(x_2)$. For each formula ϕ with $\Box\phi \in \Gamma$ we have

$$\begin{aligned} a_1 \Vdash \Box\phi \ &\Rightarrow\ x_1 \Vdash \Box\phi \\ &\Rightarrow\ x_1 \Vdash \phi \wedge \Box\phi \\ &\Rightarrow\ \begin{cases} x_1 \Vdash \phi \\ \text{and} \\ x_2 \Vdash \phi \wedge \Box\phi \end{cases} \Rightarrow \begin{cases} a_1 \Vdash \phi \\ \text{and} \\ a_2 \Vdash \phi \wedge \Box\phi \end{cases} \end{aligned}$$

where these implications follow

- using the Γ induced equivalence,

- since x_1 is reflexive,

- by the transition $x_1 \longrightarrow x_2$ and transitivity,

- using the Γ induced equivalence.

Thus we have

$$f(x_1) \longrightarrow f(x_2)$$

as required.

The remaining filtration properties are easy to verify.

Finally we need to check that $\mathcal{B}$ model KE4.

Transitivity follows in the same way as for the Lemmon filtration. For the characteristic property consider any $b_1, b_2 \in B$ with

$$b_1 \longrightarrow b_2.$$

Then, for each $a_1 \in b_1, a_2 \in b_2$, and each formula ϕ with $\Box \phi \in \Gamma$, we have

$$a_1 \Vdash \Box \phi \Rightarrow a_1 \Vdash \phi \text{ and } \cdots$$

so that

$$a_1 \Vdash \Box \phi \Rightarrow a_1 \Vdash \phi \wedge \Box \phi$$

which is enough to verify that $b_1 \longrightarrow b_1$, as required.

13.6 (a) We know that a structure $\mathcal{A}$ models KF precisely when

$$a \longrightarrow b \Rightarrow b \text{ is reflexive}$$

holds for all $a, b \in A$. Using an argument similar to that of Solution B.13 it is easy to check that this property is preserved by left-most filtrations.

(b) For a valued structure $(\mathcal{A}, \alpha)$ where $\mathcal{A}$ models KF4, consider the usual set up

$$A \xrightarrow{\ f\ } B$$

induced by a set Γ of formulas. Let $\longrightarrow$ be the transition relation on B defined by

$$b_1 \longrightarrow b_2 \Leftrightarrow \begin{cases} \text{For all } a_1 \in b_1, a_2 \in b_2 \text{ and} \\ \text{formulas } \phi \text{ with } \Box \phi \in \Gamma, \\ a_1 \Vdash \Box \phi \Rightarrow a_2 \Vdash \Box \phi \\ \text{and} \\ a_2 \Vdash \Box \phi \Rightarrow a_2 \Vdash \phi \wedge \Box \phi. \end{cases}$$

Consider the usual valuation β on this constructed transition structure $\mathcal{B}$.

We show first that f provides a morphism

$$(\mathcal{A}, \alpha) \xrightarrow{\ f\ } (\mathcal{B}, \beta).$$

Thus, consider any $x_1, x_2 \in A$ with

$$x_1 \longrightarrow x_2$$

(so that x_2 is automatically reflexive) and any $a_1 \in f(x_1), a_2 \in f(x_2)$. For each formula ϕ with $\Box \phi \in \Gamma$ we have

$$\begin{aligned} a_1 \Vdash \Box \phi &\Rightarrow x_1 \Vdash \Box \phi \\ &\Rightarrow x_1 \Vdash \Box^2 \phi \\ &\Rightarrow x_2 \Vdash \Box \phi \Rightarrow a_2 \Vdash \Box \phi \end{aligned}$$

where these implications follow

- using the Γ induced equivalence,

- by axiom 4,

- by the transition $x_1 \longrightarrow x_2$,

- using the Γ induced equivalence.

Also for each formula ϕ with $\square \phi \in \Gamma$ we have

$$a_2 \Vdash \square \phi \Rightarrow x_2 \Vdash \square \phi \Rightarrow x_2 \Vdash \phi \Rightarrow a_2 \Vdash \phi$$

where these implications follow

- using the Γ induced equivalence,

- since x_2 is reflexive,

- using the Γ induced equivalence.

Thus we have
$$f(x_1) \longrightarrow f(x_2)$$
as required.

The remaining filtration properties are easy to verify.

Finally we need to check that $\mathcal{B}$ model KE4.

Transitivity follows in the same way as for the Lemmon filtration. For the characteristic property consider any $b_1, b_2 \in B$ with

$$b_1 \longrightarrow b_2.$$

Then, for each $a_1 \in b_1, a_2 \in b_2$, and each formula ϕ with $\square \phi \in \Gamma$, we have, by the construction of $\mathcal{B}$

$$a_2 \Vdash \square \phi \Rightarrow a_2 \Vdash \phi \wedge \square \phi$$

which shows that $b_2 \longrightarrow b_2$, as required.

13.7 (a) Consider any valued structure $(\mathcal{A}, \alpha)$ where $\mathcal{A}$ is euclidean and has the E property. Consider also the usual finite set of formulas Γ. We need to construct a Γ-filtration of $(\mathcal{A}, \alpha)$ where the target is finite and has the two required properties.

Since we may work over K5 we can consider the virtual modal closure $\Gamma^{\bullet\bullet}$ of Γ, which is also finite. Consider the usual set up

$$A \xrightarrow{\;f\;} B$$

given by the $\Gamma^{\bullet\bullet}$-equivalence. Consider also the relation $\longrightarrow$ on B given by

$$b_1 \longrightarrow b_2 \;\Leftrightarrow\; \begin{cases} \text{For all } a_1 \in b_1, a_2 \in b_2 \text{ and} \\ \text{formulas } \phi \text{ with } \square\phi \in \Gamma, \\ a_1 \Vdash \square\phi \;\Rightarrow\; a_2 \Vdash \phi \\ \text{and} \\ a_1 \Vdash \square\phi \;\Rightarrow\; a_1 \Vdash \phi \\ \text{and} \\ a_1 \Vdash \Diamond\phi \;\Leftarrow\; a_2 \Vdash \phi. \end{cases}$$

This gives a structure B which we furnish with the usual valuation. We have to check the usual properties.

To verify the morphism property consider any $x_1, x_2 \in A$ with

$$x_1 \longrightarrow x_2$$

consider also $a_1 \in f(x_1), a_2 \in f(x_2)$ and a formula $\phi \in \Gamma^{\bullet\bullet}$. We know that (up to K5-equivalence) both $\square\phi, \Diamond\phi \in \Gamma^{\bullet\bullet}$. We need to check the three implications.

For the first we have

$$a_1 \Vdash \square\phi \;\Rightarrow\; x_1 \Vdash \square\phi \;\Rightarrow\; x_2 \Vdash \phi \;\Rightarrow\; a_2 \Vdash \phi$$

using the properties of the $\Gamma^{\bullet\bullet}$-induced equivalence, and the transition $x_1 \longrightarrow x_2$. For the second, since x_1 is reflexive, we have

$$a_1 \Vdash \square\phi \;\Rightarrow\; x_1 \Vdash \square\phi \;\Rightarrow\; x_1 \Vdash \phi \;\Rightarrow\; a_1 \Vdash \phi.$$

The third follows by a similar argument.

This, with some straight forward arguments, shows that f is a filtration.

Finally we need to check that B models KE5. But again this is straight forward.

13.8 (a) Since A is transitive and serial, so is B. The structure B is also finite. Consider any element b of B. Using seriality there is a chain

$$b = b_0 \longrightarrow b_1 \longrightarrow b_2 \longrightarrow \cdots$$

where, by transitivity, we have $b_r \longrightarrow b_s$ for all $r < s$. Since B is finite this chain eventually repeats, i.e. there are $r < s$ with $b_r = b_s = c$ (say). Then $b \longrightarrow c \longrightarrow c$, as required.

(b) If ϕ captures goodness, then applying (a) to $\Gamma = \Gamma(\phi)$, we have $B \Vdash^u \phi$, so that $A \Vdash^u \phi$, which need not be so.

B.14 Chapter 14

14.1 An instance of (∗3) gives

$$\vdash_{\mathsf{SLL}} \ [\bullet](\,[\bullet]\phi \to \ \square\,[\bullet]\phi\,) \to (\,[\bullet]\phi \to [\bullet]\phi\,)$$

and the rule (N) on (∗1) gives

$$\vdash_{\mathsf{SLL}} \ [\bullet](\,[\bullet]\phi \to \ \square\,[\bullet]\phi\,)$$

hence the required result follows by (MP).

14.2 (a) We show first that the assignment $a(\cdot)$ is a morphism. For each $m, n \in \mathsf{N}$ we have

$$m \longrightarrow n \ \Rightarrow \ n = m + 1$$
$$\Rightarrow \ a(n) = \mathsf{next}(a(m)) \ \Rightarrow \ a(m) \longrightarrow a(n)$$

and

$$m \xrightarrow{\ \bullet\ } n \ \Rightarrow \ (\exists r \in \mathsf{N})[n = m + r]$$
$$\Rightarrow \ (\exists r \in \mathsf{N})[a(n) = \mathsf{next}^r(a(m))] \ \Rightarrow \ a(m) \xrightarrow{\ \bullet\ } a(n)$$

which shows that $a(\cdot)$ is an unadorned morphism. Also, by construction, (Val$^{\leftrightarrow}$) holds, so that $a(\cdot)$ is a valued morphism.

It remains to check (Rel$^{\leftarrow}$).

Thus, consider any $m \in \mathsf{N}$ and $b \in \mathcal{A}$ with

$$a(m) \longrightarrow b \quad \text{or} \quad a(m) \xrightarrow{\ \bullet\ } b.$$

Then either

$$b = \mathsf{next}(a(m)) = a(m + 1) \quad \text{or} \quad b = \mathsf{next}^r(a(m)) = a(m + r)$$

for some $r \in \mathsf{N}$. This produces

$$n = m + 1 \quad \text{or} \quad n = m + r$$

with

$$m \longrightarrow n \quad \text{or} \quad m \xrightarrow{\ \bullet\ } n$$

and hence

$$a(n) = b$$

as required.

(b) The first equivalence is a direct consequence of the zigzag preservation property. For the second we have

$$\mathcal{N} \Vdash^u \phi \ \Rightarrow \ (\forall \nu)[(\mathcal{N}, \nu, 0) \Vdash^p \phi]$$
$$\Rightarrow \ (\forall \alpha, a)[(\mathcal{A}, \alpha, a) \Vdash^p \phi] \ \Rightarrow \ \mathcal{A} \Vdash^u \phi.$$

(c) Consider $A = \{a\}$, a singleton set, with the identity function on A. This gives a structure $\mathcal{A}$ with

$$\mathsf{next}(a) = a$$

and hence

$$\mathcal{A} \Vdash^u \bigcirc \phi \leftrightarrow \phi$$

for all formulas ϕ. This doesn't hold in $\mathcal{N}$.

(d) Exactly as in Theorem 14.3.

14.3 (a) Note first that if $r = i < m$ then

$$\mathsf{next}(f(r)) = \mathsf{next}(a_i) = a_{i+1} = f(r+1)$$

and if $r = m + kn + j$ where $0 \leq j < n$ then

$$\mathsf{next}(f(r)) = \mathsf{next}(b_i) = b_{i+1} = f(r+1)$$

so that

$$\mathsf{next}(f(r)) = f(r+1)$$

for all $r \in \mathsf{N}$. The required result now follows by induction.

(b) The result of (a) shows that f is a morphism. Consider any $r \in \mathsf{N}$ and $y \in A$ such that

$$f(r) \longrightarrow y \quad \text{or} \quad f(r) \overset{*}{\longrightarrow} y.$$

We need to produce some $s \in \mathsf{N}$ such that $f(s) = y$ and

$$r \longrightarrow s \quad \text{or} \quad r \overset{*}{\longrightarrow} s.$$

respectively. To do this simply consider the various possibilities as follows.

$$
\begin{array}{llll}
(1) & r < m & , \; y = a_i & \text{(some } i) \\
(2) & r < m & , \; y = b_j & \text{(some } j) \\
(3) & r \geq m & , \; y = a_i & \text{(some } i) \\
(4) & r \geq m & , \; y = b_j & \text{(some } j)
\end{array}
$$

For instance, if

$$f(r) \longrightarrow y$$

and (3) holds, then we must have $y = a_m$ $(i = m)$ and $r = m + kn + n - 1$ for some k. We then take $s = m + (k+1)n$. Similarly, if (2) holds, then we set $s = m + j$.

(c) Suppose first that $\mathcal{N} \Vdash^u \phi$. Consider any spoon $\mathcal{A}$ and valuation α on $\mathcal{A}$. Consider also the p-morphism

$$\mathcal{N} \overset{f}{\longrightarrow} \mathcal{A}$$

constructed above, and let ν be the valuation on $\mathcal{N}$ given by

$$m \in \nu(P) \iff f(m) \in \alpha(P)$$

(for all variables P). Then, by construction, f is a zigzag morphism. But $(\mathcal{N}, \nu, r) \Vdash \phi$ for all $r \in \mathbb{N}$, hence $(\mathcal{A}, \alpha, f(r)) \Vdash \phi$ so that $(\mathcal{A}, \alpha) \Vdash^v \phi$. Since α is arbitrary, this shows that $\mathcal{A} \Vdash^u \phi$.

Conversely, suppose that $\neg[\mathcal{N} \Vdash^u \phi]$, so there is some valuation ν on $\mathcal{N}$ and $r \in \mathbb{N}$ with

$$(\mathcal{N}, \nu, r) \Vdash \neg\phi.$$

By replacing $\neg\phi$ by $\langle\bullet\rangle\neg\phi$ or $\bigcirc^r \neg\phi$ we may assume that $r = 0$. Now take the modified valuation μ with

$$(\mathcal{N}, \mu, 0) \Vdash \neg\phi.$$

Consider the (m, n)-spoon $\mathcal{A}$ with the canonical p-morphism f. The choice of μ allows us to define a valuation α on $\mathcal{A}$ by

$$f(r) \in \alpha(P) \iff r \in \mu(P)$$

(for all $r \in \mathbb{N}$ and variables P from ϕ). But then f is a zigzag morphism

$$(\mathcal{N}, \mu) \xrightarrow{f} (\mathcal{A}, \alpha)$$

for this set of variables, so that $(\mathcal{A}, \alpha, f(0)) \Vdash \neg\phi$ and hence $\neg[\mathcal{A} \Vdash^u \phi]$ as required.

14.4 (a) By $(*1)$ we have $[\bullet]^2 X \subseteq [\bullet]X$. By $(*2)$ we have

$$[\bullet]X \to \square[\bullet]X = A$$

so that

$$[\bullet]([\bullet]X \to \square[\bullet]X) = A$$

and hence $(*3)$ gives $[\bullet]X \subseteq [\bullet]^2 X$, as required.

(b) Suppose $[i]$ and $[j]$ are both S-companions of $\square$. By $(*1)$ applied to $[i]$ we have

$$[i]X \to \square[i]X = A$$

so that

$$[j]([i]X \to \square[i]X) = A$$

and hence $(*3)$ and $(*1)$ applied to $[j]$ give

$$[i]X \subseteq [j][i]X \subseteq [i]X.$$

Similarly, $[i]X \subseteq [j]X$, so that $[i] = [j]$.

(c) These follow by the monotone and $\cap$-preserving properties of $\square$.

(d) Let $\mathcal{Y}$ be the set of all $Y \subseteq X$ with $DY = Y$. Note that $\emptyset \in \mathcal{Y}$. Consider $Y = \bigcup \mathcal{Y}$. For each $Z \in \mathcal{Y}$ we have $Z \subseteq Y$, so that $Z = DZ \subseteq DY$, and hence $Y = \bigcup \mathcal{Y} \subseteq DY$. This shows that $Y \in \mathcal{Y}$.

(e) A simple calculation shows that $[\bullet]$ is a box operation. Also

$$[\bullet]X = \Box [\bullet]X = [\bullet]X \cap \Box [\bullet]X \subseteq X \cap \Box [\bullet]X$$

which gives axioms $(*1, 2)$.

For axiom $(*3)$ consider

$$Y = [\bullet](X \to \Box X).$$

Then

$$Y \cap X \subseteq (X \to \Box X) \cap X \subseteq \Box X.$$

By construction we have $DY = Y$, and hence

$$D(Y \cap X) = DY \cap DX = Y \cap X \cap \Box X = Y \cap X$$

so that, (since $Y \cap X \subseteq X$), we have

$$Y \cap X \subseteq [\bullet]X$$

which gives

$$Y \subseteq (X \to [\bullet]X)$$

as required.

(f) A simple induction shows that

$$[\bullet]X \subseteq D^\alpha X$$

for all $\alpha \in Ord$. For some sufficiently large α we have $D^{\alpha+1}X = D^\alpha X$, then, with $\infty = \alpha$, we have

$$D^\beta X = D^\infty X$$

for all $\beta \geq \infty$, so that $D^\infty X = [\bullet]X$.

(g) Suppose that $\Box$ is obtained from the transition relation $\longrightarrow$. For each $r < \omega$ let

$$x \prec_r a \quad \text{mean} \quad \left\{ \begin{array}{l} \text{There is some sequence} \\ a = a_0 \longrightarrow a_1 \longrightarrow \cdots \longrightarrow a_r = x. \end{array} \right.$$

We then find that

$$a \in \Box^r X \quad \Leftrightarrow \quad (\forall x \prec_r a)[x \in X].$$

Also

$$D^r X = X \cap \Box X \cap \cdots \cap \Box^r X$$

so that

$$a \in D^\omega X \iff (\forall r < \omega)(\forall x \prec_r a)[x \in X] \iff (\forall x \prec *a)[x \in X]$$

which gives the required result.

14.5 (a) Suppose that $\mathcal{A}$ models

$$[\bullet]\Diamond P \to \langle \bullet \rangle \Box P$$

and, for $a \in A$, let

$$X = \{x \in A \mid a \xrightarrow{\bullet} x\}.$$

We wish to show that

$$(\exists x \in X)(\forall y \in X)[x \longrightarrow y \Rightarrow x = y].$$

If this doesn't hold then, modifying the argument of Solution to 5.8(b), there is a partition Y, Z of X with

$$(\forall y \in Y)(\exists z \in Z)[y \longrightarrow z] \quad \text{and} \quad (\forall z \in Z)(\exists y \in Y)[z \longrightarrow y].$$

Consider any valuation on $\mathcal{A}$ such that for each $x \in X$

$$x \Vdash P \iff x \in Y \quad , \quad x \Vdash \neg P \iff x \in Z.$$

Then

$$y \Vdash \Diamond \neg P \quad , \quad z \Vdash \Diamond P$$

for each $y \in Y$ and $z \in Z$, so that

$$x \Vdash (P \to \Diamond \neg P) \quad , \quad x \Vdash (\neg P \to \Diamond P)$$

for each $x \in X$. Since $\longrightarrow$ is reflexive (and hence $\phi \to \Diamond \phi$ holds in $\mathcal{A}$) this gives

$$x \Vdash \Diamond P \wedge \Diamond \neg P$$

for all $x \in X$, and hence

$$a \Vdash [\bullet]\Diamond P \wedge [\bullet]\Diamond \neg P$$

which is the required contradiction.

The converse is easy.

(b) If $\mathcal{A} = (A, \longrightarrow)$ is reflexive and transitive then $\longrightarrow$ and $\xrightarrow{*}$ agree and the formula considered reduces to the McKinsey formula. However, Exercise 5.8 applies to all (not necessarily reflexive) transitive structures.

B.15   Chapter 15

15.1 (a) The rule (EN) gives

$$\vdash_\mathsf{K} \ \Box S(\phi) \to \Box(\Box\phi \to \phi)$$

and hence

$$\vdash_\mathsf{LL} \ \Box S(\phi) \to \Box\phi$$

follows immediately.

(b) Consider any characteristic valued model $(\mathcal{A}, \alpha)$ for S, and let $(\mathcal{B}, \beta)$ be the structure as constructed in section 15.3. It suffices to show that

$$(\mathcal{B}, \beta, b) \ \Vdash \ \Box S(\phi) \to \Box\phi$$

for then $(\mathcal{B}, \beta)$ models S.

To this end suppose

$$b \Vdash \ \Box S(\phi).$$

Then

$$b \Vdash S(\phi) \quad \text{and hence} \quad a \Vdash \ \Box\phi \to \phi$$

for all $a \in A$. In particular

$$a \Vdash \ \Box S(\phi)$$

for all $a \in A$, so that, since $(\mathcal{A}, \alpha)$ models S, we have $a \Vdash \phi$. This gives

$$b \Vdash \ \Box\phi$$

as required.

(c) Exactly as in Lemma 15.5.

(d) No. Indeed, with

$$S(\phi) := \phi$$

we see that S is just K, which is canonical.

15.2 (a) (i) The displayed formula is an instance of a tautology.
(ii) Since $\psi \to \phi$ is a tautology, this follows by (EN).
(iii) Set $\xi := \Box\psi$ in (i) and use (ii).
(iv) By (iii), (EN), and K.
(v) By (iv).

(b) Using (EN) on (v) we have

$$\vdash_\mathsf{K} \ \Box\phi \to \Box U(\psi) \quad \text{and hence} \quad \vdash_\mathsf{S} \ \Box\phi \to \psi$$

which gives the two required results.

15.3 (a) By the previous Exercise we have

$$\vdash_{\mathsf{KG}} T(\phi) \quad , \quad \vdash_{\mathsf{KG}} 4(\phi).$$

The second of these gives

$$\vdash_{\mathsf{KG}} \square U(\phi) \to \square^2 U(\phi).$$

Also, an application of (EN) to Grz(ϕ) gives

$$\vdash_{\mathsf{KG}} \square^2 U(\phi) \to \square\phi$$

so that

$$\vdash_{\mathsf{KG}} Hrz(\phi).$$

This shows that $\mathsf{KHT} \leq \mathsf{KG}$.

The converse, $\mathsf{KG} \leq \mathsf{KHT}$, is immediate.

(b) Consider any characteristic valued model $(\mathcal{A}, \alpha)$ of KH and let $(\mathcal{B}, \beta)$ be the rooted structure as constructed in Section 15.3. It suffices to show that

$$b \Vdash H(\phi)$$

for then $(\mathcal{B}, \beta)$ models KH.

To this end suppose that

$$b \Vdash \square U(\phi).$$

Then $\mathcal{A} \Vdash U(\phi)$ and in particular

$$\mathcal{A} \Vdash \square U(\phi)$$

for all $a \in A$. Since $(\mathcal{A}, \alpha)$ models KH, this gives $a \Vdash \square\phi$, for all a, and in particular $a \Vdash \square^2\phi$. But this gives

$$a \Vdash \square(\phi \to \square\phi)$$

so that, using U(ϕ), we have $a \Vdash \phi$, and hence $b \Vdash \square\phi$, as required.

(c) As usual, consider any characteristic model $(\mathcal{A}, \alpha)$ of KG. Let $(\mathcal{B}, \beta)$ be the rooted structure as constructed in Section 15.3 but modified to make the root b reflexive. (This means that equivalence (15.6) is no longer automatic.) It suffices to show that

$$b \Vdash Grz(\phi).$$

To this end, suppose that

$$b \Vdash \square U(\phi).$$

Then

$$b \Vdash U(\phi) \quad \text{and} \quad a \Vdash U(\phi)$$

for all $a \in A$. Since $(\mathcal{A}, \alpha)$ models **KG**, the second of these gives

$$a \Vdash \phi \quad , \quad a \Vdash \square \phi$$

for all $a \in A$. Thus, remembering that

$$b \longrightarrow x \Leftrightarrow x = b \text{ or } x \in A$$

we have

$$b \Vdash \phi \rightarrow \square \phi.$$

Also,

$$a \Vdash \phi \rightarrow \square \phi.$$

for all $a \in A$, so that

$$b \Vdash \square(\phi \rightarrow \square \phi).$$

Since $b \Vdash U(\phi)$, this gives $b \Vdash \phi$, as required to complete the argument.

(d) Let $\phi := \square P \rightarrow P$. We know that $\vdash_{\mathsf{KG}} \phi$, so it suffices to produce a non-reflexive structure which models **KH**.

Let $\mathcal{N} = (\mathsf{N}, \longrightarrow)$ where

$$m \longrightarrow n \Leftrightarrow n < m.$$

This is irreflexive. Consider any valuation on $\mathcal{N}$ and $m \in \mathsf{N}$ with

$$m \Vdash \square U(\phi).$$

We require that $m \Vdash \square \phi$.

If this doesn't hold then there is some $n < m$ with $n \Vdash \neg \phi$. We take the least such n and derive a contradiction.

Since $n < m$ we have $n \Vdash U(\phi)$ so that

$$n \Vdash \neg \square(\phi \rightarrow \square \phi)$$

(otherwise $n \Vdash \phi$). This gives us some $l < k < n$ with

$$k \Vdash \phi \wedge \Diamond \neg \phi \quad , \quad l \Vdash \neg \phi$$

which contradicts the minimality of n.

B.16   Chapter 16

16.1 For each pair of formulas ϕ and ψ, the axioms 4, K, and T give

$$\vdash_{\mathsf{S4}} \square \phi \rightarrow \square^2 \phi \quad , \quad \vdash_{\mathsf{S4}} \square^2 \phi \rightarrow \square^3 \phi$$

$$\vdash_{\mathsf{S4}} \square(\square^2 \phi \rightarrow \square \psi) \rightarrow (\square^3 \phi \rightarrow \square^2 \psi)$$

$$\vdash_{\mathsf{S4}} \square^2 \psi \rightarrow \square \psi \quad , \quad \vdash_{\mathsf{S4}} \square^2 \psi \rightarrow \square \psi$$

and hence a tautology gives $\vdash_{\mathsf{S4}} (*)$.

16.2 (a) $R(i, j, k, l)$ is an instance of $S(i, j, k, l)$.

(b) Using the T axiom, the given comparisons ensure that for all formulas θ, ϕ, ψ we have

$$\vdash_{\mathsf{KT}} \square^i \theta \to \square^{i'} \theta \quad , \quad \vdash_{\mathsf{KT}} \square^{j'} \phi \to \square^j \phi$$

$$\vdash_{\mathsf{KT}} \square^k \psi \to \square^{k'} \psi \quad , \quad \vdash_{\mathsf{KT}} \square^{l'} \theta \to \square^l \theta$$

so that

$$\vdash_{\mathsf{KT}} (\square^j \phi \to \square^k \psi) \to (\square^{j'} \phi \to \square^{k'} \psi)$$

and hence, eventually

$$\vdash_{\mathsf{KT}} S(i', j', k', l') \to S(i, j, k, l).$$

(c) Let $\mathcal{A}$ be a finite model of $R(i, j, k, l)$. Then, since $\square$ is deflationary, for each $X \subseteq A$ we have a descending chain

$$X \supseteq \square X \supseteq \square^2 X \supseteq \ldots \supseteq \square^r X \supseteq \cdots$$

which (since A is finite) must eventually stabilize. Observe that we have

$$\square^j X = \square^k X \;\Rightarrow\; \square X = \square^2 X.$$

Consider any $r \in \mathsf{N}$ with

$$\square^{r+2} X = \square^{r+3} X.$$

Then

$$\square^{r+2} X = \square^{r+3} X = \square^{r+4} X = \cdots$$

in particular

$$\square^{r+j} X = \square^{r+k} X$$

so that

$$\square^{r+1} X = \square^{r+2} X.$$

Thus, by induction

$$\square^{r+1} X = \square^{r+2} X \;\Rightarrow\; \square X = \square^2 X$$

as required.

(d) This follows from Exercise 10.3.

(e) Using the structural property for $K(i, j, k, l, 1, 2)$ we need to show that for each $a, b, c \in \mathsf{N}$ with

$$a \xrightarrow{\ l\ } b \xrightarrow{\ 2\ } c$$

there is some $d \in \mathbb{N}$ with

$$a \xrightarrow{\,i\,} d \xrightarrow{\,k\,} c \quad , \quad d \xrightarrow{\,j\,} x \to b \longrightarrow x$$

(for each $x \in \mathbb{N}$). Here the affix on the arrows indicates the number of steps.
For $x, y \in \mathbb{N}$ we have

$$x \xrightarrow{\,r\,} y \iff x \le y + r$$

so we have

$$a \le b + l \quad , \quad b \le c + 2.$$

Since $2 \le j$ we may set

$$d = b + j - 1.$$

Then, since $l + 1 \le i + j$, we have

$$a \le b + l \le b + i + j - 1 = d + i$$

so that

$$a \xrightarrow{\,i\,} d,$$

and, since $j < k$, we have

$$d = b + j - 1 \le c + j + 1 \le c + k$$

so that

$$d \xrightarrow{\,k\,} c.$$

Finally

$$d \xrightarrow{\,j\,} x \;\Rightarrow\; d \le x + j \;\Rightarrow\; b \le x + 1 \to b \longrightarrow x$$

as required.

(f) All finite models of $\mathsf{R}(i, j, k, l)$ are transitive, but $\mathsf{S}(i, j, k, l)$ has a non-transitive infinite model.

B.17   Chapter 17

17.1 (a)(i) By induction we find that $[\![\,\Box^{r+1}\bot\,]\!] = \{0, \ldots, r\}$.

(ii) $[\![\,\Box\Diamond\bot\,]\!] = \Box\emptyset\{0\}$

(iii, iv) The two computations of similar.

$$
\begin{aligned}
[\![\,\Box\Diamond\Box\top\,]\!] \qquad\qquad\quad & [\![\,\Box\Diamond\Box\bot\,]\!] \\
= \;\Box\Diamond a \qquad\qquad\qquad & = \;\Box\neg\Box\emptyset \\
= \;\Box\Diamond\{0\} \qquad\qquad\quad & = \;\Box\neg\{0\} \\
= \;\Box\neg\Box\{1, \ldots, \infty\} \qquad & = \;\Box\{1, \ldots, \infty\} \\
= \;\Box\{1, \ldots, \omega\} \qquad\qquad & = \;\{0, \infty\} \\
= \;\{0, \infty\} \qquad\qquad\qquad &
\end{aligned}
$$

(v) For all θ we have

$$0 \in [\![\,\square\,\theta\,]\!] \quad \text{and} \quad \infty \in [\![\,\square\,\theta\,]\!] \Leftrightarrow \omega \in [\![\theta]\!]$$

so that

$$[\![Z(\theta)]\!] = \begin{cases} A & \text{if } \omega \in [\![\theta]\!] \\ \neg\{\infty\} & \text{if } \omega \notin [\![\theta]\!]. \end{cases}$$

(vi) Consider $U = [\![\theta]\!]$. If $\mathsf{N} \subseteq U$ or $\omega \notin U$ then $\square U \subseteq \square^2 U$, so that

$$[\![4(\theta)]\!] = A.$$

Otherwise we have

$$\square U = \{0, \ldots, n, \infty\} \quad , \quad \square^U = \{0, \ldots, n+1\}$$

for some $n \in \mathsf{N}$, so that

$$[\![4(\theta)]\!] = \neg\{\infty\}.$$

(b) (i)A and $\mathsf{N} \cup \{\omega\}$ (ii)$\neg\{0, \infty\}$ (iii)$\emptyset$

(c) Let $U_r = \square^r\emptyset$, so that

$$U_0 = \emptyset \quad , \quad U_{r+1}\{0, \ldots, r\}.$$

With this family $\mathcal{U}$ we have $\bigcup\mathcal{U} = \mathsf{N}$, so that $\square\bigcup\mathcal{U} = \mathsf{N} \cup \{\omega\}$, whereas $\square\mathcal{U} \subseteq \mathcal{U}$, so that $\bigcup\square\mathcal{U} = \mathsf{N}$.

17.2 No. Suppose that the finite valued structure $(\mathcal{A}, \alpha)$ models KY. For each valuation μ on $\mathcal{A}$ there is a substitution $P \mapsto \mu(P)$ such that

$$(\mathcal{A}, \mu) \Vdash^v \phi \Leftrightarrow (\mathcal{A}, \alpha) \Vdash^v \phi^\mu$$

holds for all formulas ϕ. Since $Y(\phi)^\mu = Y(\phi^\mu)$ this means that

$$(\mathcal{A}, \alpha) \Vdash^v Y(\phi^\mu) \quad , \quad (\mathcal{A}, \alpha) \Vdash^v Y(\phi)^\mu \quad , \quad (\mathcal{A}, \mu) \Vdash^v Y(\phi)$$

hold so that $\mathcal{A} \Vdash^u Y(\phi)$. Thus $\mathcal{A}$ models KY and hence also models KZ.

17.3 (a) Consider any temporal structure $\mathcal{A} = (A, \xrightarrow{+}, \xrightarrow{-})$ (so that $\xrightarrow{+}, \xrightarrow{-}$ are converse transitive relations). Suppose that $\mathcal{A}$ models M_+. Then, by Exercise 5.8, for each $a \in A$ there is some $b \in A$ with $a \xrightarrow{+} b$ and such that

$$b \xrightarrow{+} x \Rightarrow x = b$$

(for all $x \in A$). Take such a $b \in A$. The first part of the property, this time applied to b, gives some c with $b \xrightarrow{+} c$, and hence $b \xrightarrow{+} b$. But now we have an infinite chain

$$b \xrightarrow{-} b \xrightarrow{-} b \xrightarrow{-} \cdots$$

which, by Theorem 5.7, obstructs $\mathcal{A}$ from modelling L_-.

(b)(i) $\mathcal{N}$ is transitive and well founded in the $(-)$-direction.

(ii) For each $X \subseteq \mathsf{N}$ and $a \in A$ we have

$$a \in [+]\langle+\rangle x \;\Leftrightarrow\; (\forall y > a)(\exists x > y)[x \in X].$$

In particular, if $[+]\langle+\rangle X$ is non-empty then X can not have a maximum member, and so X is infinite. But then the right hand condition is satisfied by any $a \in \mathsf{N}$.

(iii) Let $\mathcal{X}$ be the family of finite/cofinite subsets of N. Modifying the previous argument we see that

$$\begin{aligned} X \text{ finite} \;&\Rightarrow\; [+]X = \emptyset \\ X \text{ cofinite} \;&\Rightarrow\; [+]X = \text{cofinite}. \end{aligned}$$

Thus $\mathcal{X}$ is closed under the boolean operations and the box operation $[+]$. This shows that $\llbracket \phi \rrbracket_\nu \in \mathcal{X}$ for all formulas ϕ. Finally, with $X = \llbracket \phi \rrbracket_\nu$, one of $X, \neg X$ is finite and the other is cofinite so, by (ii), one of

$$[+]\langle+\rangle X \quad , \quad [+]\langle+\rangle\neg X$$

is $\emptyset$ and the other is N. Thus

$$[+]\langle+\rangle X \;=\; \langle+\rangle[+]X$$

and hence $\llbracket \mathrm{M}(\phi) \rrbracket_\nu = \mathsf{N}$.

(c) If S is Kripke-complete then, by (a), we have $\vdash_\mathsf{S} \bot$. This is prevented by the existence of at least one valued model.

Bibliography

[1] S. Abramsky, D. Gabbay, and T. Maibaum. *Handbook of Logic in Computer Science,*. Oxford University Press, 1992. (Vols 1 and 2, further volumes to be published).

[2] S. Abramsky and S. Vickers. Quantales, observation logic and process semantics. *Mathematical Structures in Computer Science*, 3, 1993. (pp. 161 – 227).

[3] J. W. de Bakker, W. P. de Roever, and G. Rozenberg. *Linear Time, Branching Time and Partial Order in Logics and Models for Concurrency*, volume 350 of *Springer Lecture Notes in Computer Science*. Springer-Verlag, 1989.

[4] B. Banieqbal and H. Barringer. Temporal logic with fixed points. In *Temporal Logic in Specification*. Springer-Verlag, 1989. (pp. 62 – 74 of [5]).

[5] B. Banieqbal, H. Barringer, and A. Pnueli. *Temporal Logic in Specification*, volume 398 of *Springer Lecture Notes in Computer Science*. Springer-Verlag, 1989.

[6] H. Barringer and R. Kuiper. Temporal logic and concurrency. In *Handbook of Logic in Computer Science*. Oxford University Press, 199? (to appear in a volume of [1]).

[7] J. F. A. K. van Benthem. A note on modal formulae and relational properties. *J. Symbolic Logic*, 40:55 – 58, 1975.

[8] J. F. A. K. van Benthem. Modal reduction principles. *J. Symbolic Logic*, 41:301 – 312, 1976.

[9] J. F. A. K. van Benthem. *The logic of time*. Reidel, 1982.

[10] J. F. A. K. van Benthem. *Modal Logic and Classical Logic*. Bibiliopolis, Naples, 1983.

[11] J. F. A. K. van Benthem. Correspondence theory. In *Handbook of Philosophical Logic, vol. II*. Reidel, 1984. (pp. 167 – 247 of [24]).

311

[12] J. F. A. K. van Benthem. Time, logic and computation. In *Linear Time, Branching Time and Partial Order in Logics and Models for Concurrency*, volume 350 of *Springer Lecture Notes in Computer Science*. Springer-Verlag, 1989. (pp. 1 – 49 of [3]).

[13] J. F. A. K. van Benthem. Temporal logic. In *Handbook of Logic in Artificial Intelligence and Logic Programming*. Oxford University Press, 199? (to appear in a volume of [25]).

[14] G. Boolos. *The unprovability of consistency*. Cambridge University Press, 1979.

[15] R. Bull and K. Segerberg. Basic modal logic. In *Handbook of Philosophical Logic, vol II*. D. Reidel and Co., 1984. (pp. 1 – 88 of [24]).

[16] J. P. Burgess. Basic tense logic. In *Handbook of Philosophical Logic, vol II*. D. Reidel and Co., 1984. (pp. 89 – 133 of [24]).

[17] B.F. Chellas. *Modal Logic, an introduction*. Cambridge University Press, 1980.

[18] E. A. Emerson. Temporal and modal logic. In *Handbook of Theoretical Computer Science, vol B*. Elsevier, 1990. (Chapter 16, pp. 995 – 1072 of [35]).

[19] E. A. Emerson and J. Srinvasan. Branching time temporal logic. In *Linear Time, Branching Time and Partial Order in Logics and Models for Concurrency*, volume 350 of *Springer Lecture Notes in Computer Science*. Springer-Verlag, 1989. (pp. 123 – 284 of [3]).

[20] F. B. Fitch. A correlation between modal reduction principles and properties of relations. *J. Philosophical Logic*, 2:97 – 101, 1973.

[21] M. Fitting. *Proof Methods for Modal and Tense Logics*. Reidel, 1983.

[22] M. Fitting. Basic modal and temporal logics. In *Handbook of Logic in Artificial Intelligence and Logic Programming*. Oxford University Press, 1993. (A chapter of volume 1 of [25]).

[23] D. Gabbay. *Investigations into modal and tense logics, with applications to problems in linguistics and philosophy*. Reidel, 1976.

[24] D. Gabbay and F. Guenther, editors. *Handbook of Philosophical Logic*, volume II. D. Reidel and Co., 1984.

[25] D. Gabbay, C. J. Hogger, and J. A. Robinson, editors. *Handbook of Logic in Artificial Intelligence and Logic Programming*. Oxford University Press, 1993. (Vol 1, further volumes to appear).

[26] A. Galton, editor. *Temporal logics and their applications*. Academic Press, 1987.

[27] R. Goldblatt. *Logics of time and computation*. Number 7 in Lecture Notes. CSLI, 1987.

[28] D. Harel. Dynamic logic. In *Handbook of Philosophical Logic, vol II*. D. Reidel and Co., 1984. (pp. 497 – 604 of [24]).

[29] G. E. Hughes and M. J. Cresswell. *An introduction to Modal Logic*. Methuen, 1972.

[30] G. E. Hughes and M. J. Cresswell. *A companion to Modal Logic*. Methuen, 1984.

[31] H. Kamp. *Tense Logic and the Theory of Linear Order*. PhD thesis, University of California at Los Angeles, 1968.

[32] S. Kanger, editor. *Proceedings of the 3rd Scandinavian Logic Symposium*. North-Holland, 1975.

[33] K. Konolige. Autoepistemic logic. In *Handbook of Logic in Artificial Intelligence and Logic Programming*. Oxford University Press, 199? (to appear in a volume of [25]).

[34] S. Kripke. Semantic analysis of modal logic I. Normal modal propositional calculi. *Zietshrift f. math. Log. u. Grund. math*, 9:67 – 96, 1963.

[35] J. van Leeuwen. *Handbook of Theoretical Computer Science, vol B: Formal Models and Semantics*. Elsevier, 1990.

[36] E. J. Lemmon. Algebraic semantics for modal logics I. *J. Symbolic Logic*, 31:46 – 65, 1966.

[37] E. J. Lemmon. Algebraic semantics for modal logics II. *J. Symbolic Logic*, 31:191 – 218, 1966.

[38] E. J. Lemmon and D. S. Scott. *The 'Lemmon' notes: An Introduction to Modal Logic*. Blackwell, 1977.

[39] A. Macintyre and H. Simmons. Gödel's diagonalization technique and related properties of theories. *Colloq. math.*, 28:165 – 180, 1973.

[40] Z. Manna and A. Pnueli. *The Temporal Logic of Reactive and Concurrent Systems*. Springer-Verlag, 1991.

[41] H. Sahlqvist. Completeness and correspondence in the first- and second-order semantics for modal logic. In *Proceedings of the 3rd Scandinavian Logic Symposium*. North-Holland, 1975. (pp. 110 – 143 of [32]).

[42] K. Segerberg. *An essay in Classical Modal Logic*. Department of Philosophy, University of Uppsala, 1971.

[43] C. Smoryński. Modal logic and self-reference. In *Handbook of Philosophical Logic, vol II*. D. Reidel and Co., 1984. (pp. 441 – 495 of [24]).

[44] C. Smoryński. *Self-reference and modal logic*. Springer-Verlag, 1985.

[45] R. Smullyan. *Forever Undecided, A puzzle guide to Gödel*. Oxford University Press, 1988.

[46] C. Stirling. Temporal logics for CCS. In *Linear Time, Branching Time and Partial Order in Logics and Models for Concurrency*, volume 350 of *Springer Lecture Notes in Computer Science*. Springer-Verlag, 1989. (pp. 660 – 672 of [3]).

[47] C. Stirling. Comparing linear and branching time temporal logics. In *Temporal Logic in Specification*, volume 398 of *Springer Lecture Notes in Computer Science*. Springer-Verlag, 1989. (pp. 1 – 20 of [5]).

[48] C. Stirling. Modal and temporal logics. In *Handbook of Logic in Computer Science*. Oxford University Press, 1992. (pp. 477 – 563 of volume 2 of [1]).

[49] R. H. Thomason. Combinations of tense and modality. In *Handbook of Philosophical Logic, vol II*. D. Reidel and Co., 1984. (pp. 135 – 165 of [24]).

[50] L. A. Wallen. *Automated Deduction in nonclassical logics*. The MIT Press, 1990.